The Human Computer

The Human Computer

✦

Get The Most Out Of Yours!

EARTH PRO Series

Dr. Anthony Scheiber

Writers Club Press

New York Lincoln Shanghai

The Human Computer
Get The Most Out Of Yours!

Writers Club Press
an imprint of iUniverse, Inc.

For information address:
iUniverse, Inc.
2021 Pine Lake Road, Suite 100
Lincoln, NE 68512
www.iuniverse.com

ISBN: 0-595-25293-1 (pbk)
ISBN: 0-595-65076-7 (cloth)

Printed in the United States of America

Illustrations and cover layout: Anthony Scheiber

Thanks to Karin and Bryan, CJ & Kevin
for their patience and encouragement,
&
my Father and Mother,
& Pat
& Stefan

for all their help and support.

"The door that exits between reality and fiction may be as fragile as how accepting our mind is in embracing change to our current, pretentious paradigms."

—AS

Contents

A *THE PERFECT CRIME*

I. *FINALLY, THE TRUTH ABOUT YOUR LIFE*

II. *UNDERSTANDING COMPUTERS*

III. HUMANS AS COMPUTERS

B *THE PERFECT IDENTITY SWITCH*

IV. ESSENTIAL HUMAN COMPUTER FUNCTIONS

V. STRETCHING YOUR IMAGINATION EVEN FURTHER

Δ INVESTIGATING THE PERFECT CRIME

VI. WELL BEYOND THE REACH OF EVERY DAY FICTION

Φ PULLING OFF THE PERFECT CRIME

Ω *THE PERFECT ENDING TO THE PERFECT CRIME*

PROLOGUE:
WHY YOU <u>NEED</u> TO READ
THIS BOOK

Because you <u>need</u> to read this book…No, not because someone simply suggested you should read this book, but to keep alive the quest for newer, better technology, ideas, concepts, in a world where our current digital technology paradigms are rapidly headed towards extinction, *you* have to read this book. To move your deepest personal relationships forward, *you* should read this book. To propel humanity forward to a greater degree of enlightenment, *you* might read this book. To understand how you and the world around you interconnect, *you* really have to read this book. Because this book will change the world you live in and *you* should be the first to know why. *But the most important reason to read this book,*…to understand yourself better, to peer inside your head and know how your brain functions, to understand better how you work, who you are and how incredibly successful you, yes you, can be…*you must* read this book.

Humans are great mimickers. Over the centuries architects and engineers have had great respect for Mother Nature and her designs, both in animal life forms as well as natural physical wonders. Often human architects, engineers, artists, and everyday people design and build things modeled after Nature's exquisite designs. And why not?…Mother nature has had 4.6 billion years to perfect her designs.

Understanding how the human brain functions may save us. Technologically speaking, we are in a race against a time clock that is steadily ticking. The last century saw an explosion of technologic advances. The first commercial vacuum tube was created by John Ambrose Fleming in 1904. World war II instigated the building of

ENIAC (Electronic Numerical Integrator Analyzer and Computer) comprised of 17,468 vacuum tubes and designed to produce missile trajectory tables. In 1926, the first semiconductor transistor was invented, but it took until 1947, for Bell Labs' William Shockley to introduce the modern solid-state, reliable transistor. The year 1958, saw the first integrated circuit built by Texas Instruments. IBM's 7000 series mainframes, in 1959, represented the company's first transistorized computers. In 1975, the Altair 8800, made by MITS, appeared as the forerunner of the personal computer. The last twenty-seven years the computer has undergone a tremendous metamorphosis, changing size, shape, speed, processing power, memory capacity, interconnections with internal and external devices and ever expanding and extending its usefulness. The colossal wave of technology is built on a single, limiting, choking premise,…that of binary (digital) technology…it's either 'on' or it's 'off'. Technology as we know it today is doomed…unless it evolves in a radically new way…and fast.

A schoolmate in high school once broke up with a girlfriend by stating a common phrase at the time, "Relationships are like a shark; sharks have to keep moving ahead to stay alive; and what we have is a dead shark." Technology as we know it, must not only keep forging ahead, but continuously explore new waters, if it is to stay alive.

In a physical sense we can only make switches that go 'on' and 'off' so small…a couple of atoms might do, but at some point restricting the storage and flow of information to a 'zero' or a 'one' is a dead end street. It has been suggested that within twenty years, digital technology will hit its wall. It is time to seek out a set of new fundamentals upon which to build future paradigms, to look for a better manner by which to handle and process information other than with digital technology as we know it today.

Unless an alien species pays us a visit and offers to relinquish the blueprints to an entirely new computer technology, or an as of yet unidentified warrior computer genius rises up from the population and leads us on a crusade to radically change our current technological paradigms,...we must aggressively look outside the box, before we can get out of the box. Learning more about the human brain, the most sophisticated computer on the planet, and how it functions, may be our only way to get out of the box.

Investigating how the human brain is constructed, how it functions, how it stores data and processes information may lead to important clues to spur on the evolution of computer technology. We will create an entirely new discipline: *The pioneer age of neurocomputerology.*

On the other hand, the personal computer, one of mankind's greatest achievements, may up until now have been simply be a reflection of ourselves. Understanding the personal computer, how it is constructed and functions, may provide us with important clues as to how the human brain works.

The principles of how our human computer functions may also provide the tools for solving some of life's greatest mysteries. We may better understand such interests as the drive behind sexual attraction, and the ultimate purpose behind the most powerful force in the universe *true love.* Knowing how the human computer works also provides one with key insight into our individual destinies and the vast potential each of us possess. A successful, rewarding, financially prosperous life may be within everyone's grasp regardless of one's age, and simply be a matter of understanding how to explore one's mind and exploit the treasures locked away in ancient memory files located in the yet unexplored crevices of our brains.

In the following pages, we will investigate the principles behind the desktop computer, illustrate the anatomy of the human brain, explore how the human brain functions, draw parallels between the human brain and the desktop computer, then finally uncover life's secrets locked away in the human computer. This is neither a medical or a technical manual. This book represents the first attempt to translate and amalgamate two very sophisticated bodies of information into one resource. An ambitious proposal, but well worth your time.

A GUIDE TO READING THIS BOOK

If you are like me, I needed the whole story organized and presented in an orderly fashion for this very complex subject to make any sense. I also enjoyed trying to understand the basics of technology and the human brain, before leaping forward to understand the similarities between the computer and the human brain, to facilitate being able to make connections between these two highly technical topics.

However, I understand that there are many readers that already have a comprehensive understanding of electronics and computers. If you are comfortable in your knowledge of how the desktop computer is constructed and how it works, then by all means, skip the first nine chapters and start your reading with 'Anatomy of the Human Brain'-Chapter Ten.

If, on the other hand, you are a reader not interested in how either a computer is constructed and works, or how the human brain is constructed, but you are interested in seeking the valuable secrets of this book, skip to Chapter Twenty-Eight 'Why I don't understand my teen', and start your reading there.

If you are <u>most</u> interested in the why's and how's of the universal struggle of Women versus Men, and the reason behind how we pick a 'soulmate', skip straight to Chapter Twenty-Nine, but then jump to Chapter Thirty-Eight, then Chapter Thirty-Nine and Forty (Secret Behind the Most Powerful Force in the Universe). I believe the world of *Romance Novels* will all start to make much more sense to you, and so will your past and future relationships.

Oh, you don't want to miss the short story scattered across the book... Enjoy.

A
THE PERFECT CRIME

Nevin Thomas stood with a rubber mask on his face. His right hand clutched the smooth wooden handle of a chilling, cold steel 44 magnum, laser targeted pistol. The face on the rubber mask he wore was that of George Washington. Brilliant emerald green globes stared through the eye slits.

Nevin Thomas slowly fanned the barrel of his gun at the crowd that nervously stood before him. He coughed, then stated in a gruff, firm voice, "Relax,…and no one will get hurt."

Accompanying Nevin Thomas, were two armed accomplices. The shorter bank robber wore a mask that bore the likeness of Ben Franklin, while the taller accomplice wore a mask that was the facsimile of Abraham Lincoln.

The three gunmen held an audience of sixteen men and women spellbound, swaying their gun barrels back and forth. All sixteen hostages stood silently inside the Oak Ridge National bank, their hands nervously raised towards the ceiling. The bank's guard had been neutralized with a stun gun, and lay unconscious on the floor.

The bank robbers' timing had been perfect. The vault door was open and an abundance of cash was resting at each teller's station in anticipation of a busy Friday. However, the readily available loot in the tellers' drawers was not the robbers' objective. The man who wore the Ben Franklin's mask directed the bank manager to unlock the gate that barred the entrance to the inner vault chamber.

As the bank manager struggled nervously to open the vault door, a blond twenty-one year old woman, dressed in a flashy red dress, sporting dark sun glasses on her soft, petite nose stood quietly in the crowd. She carried a stun gun and mace in her bag.

Frightened, the middle-aged bank manager, father of two, followed the robber's directions exactly. He had no intention of aggravating the masked men, and dying an early death. He pulled the key from his pocket and unlocked the metal gate. The sturdy iron gate glided open on well-oiled hinges.

Nevin and the bank robber wearing Ben Franklin's mask hustled into the inner chamber of the vault. Inside were six money sacks loaded with cash. The bank had just received a sizeable deposit of unmarked bills from a local casino. The robbers eagerly snatched up the loot and made their exit.

Like clockwork, the team exited the bank. The three trotted down the front steps into the waiting car idling on the street outside the bank. Using the electronic key, Nevin unlocked the car doors. The entire operation had taken less than four minutes. The robbers had studied the bank for three months. They had hit the bank like a precision commando team. The three men glided into the getaway car.

Seated in the back of the black sedan, Nevin Thomas ripped the mask of George Washington from his face. The smell of the rubber choked him. He popped out the green tinted contacts he wore to disguise his true eye color. The two contacts fell to the floor. There was a tickle in the back of his throat. He coughed. The lower end of his breathing tube, the trachea, hurt as he let the cough out. His lung cancer was slowly, but steadily progressing.

Using his right hand he swept his hair back. As the hand reached the back of his head, his fingers brushed against a plastic plug fixed to his scalp, hidden by the strands of his dark, chestnut brown hair.

I. FINALLY, THE TRUTH ABOUT YOUR LIFE

REALITY is that we live in a physical universe, governed by the laws of nature, physics and mathematics. As humans we are intrigued by the possibility of magic and the supernatural. We often apply superstition to that which we don't understand, that which captivates our fancy; that which some wish to exploit.

That which occurs in our life, occurs under rules governed by scientific principles. Even the supernatural must utilize the atoms and the energy that exist around us. The amount of energy by which things occur must conform to Einstein's equation, $E=MC^2$ or 'energy equals mass multiplied by mathematical value of the speed of light squared'.

So that the construction of the human race, either as the product of billions of years of evolution, or at the hand of creation-must still conform to scientific principles-because of the physical nature of our world. Whether one believes in evolution, or in creation, is really asking the question of how much energy did it take to construct the human race?-either a lot of energy or a colossal amount of energy. The mass or the physical occurrence remains the same, or in other words, the one truth that envelopes us, is the truth that it did occur, and continues to occur with each succeeding human born, and all of mankind has been forged from the raw materials that exist on the planet we live upon.

The truth about life, your existence, is that whether we are the product of some Deity's desire, or whether we are the product of a random and very lucky set of mutational events, like the evolution theorists would like us to believe, or if you have no set belief-any way you look at it, a definable mechanism has been, and is still at work making human life possible.

Whether one believes that Adam & Eve existed some twelve thousand years ago, or that the archeology carbon dating of recent skull finds suggest that the first humans walked the Earth possibly two million years ago, something methodical has to be at work to successfully produce generation after generation of relatively flawless humans. Whether it has been thousands of years or hundreds of thousands of years, we would even tend to believe that each succeeding generation of human is improving, growing taller, becoming stronger and is more intelligent.

Hippocrates-whom many consider the Father of Medicine, earned this title in 500 BC because he is thought to be the first human to recognize that illness and disease were not the result of the twisted work of evil supernatural demons, but the result of physical afflictions. While his colleagues prayed over the sick, he professed that illness could be treated. Without the aide of a microscope to prove his theory, Hippocrates championed the use of a variety of medical and surgical remedies to treat the diseases that afflicted mankind.

Hippocrates recognized the physical nature of the world that surrounded him. He chose to take advantage of what he could define and change, given the primitive tools at his disposal.

In the midst of a world that believed the Earth was the center of the universe, Galileo chose to construct a telescope and look out beyond the paradigms of his existence. His findings supported Copernicus's theory that the Earth was a planet that circled a stationary sun. Inflamed by the audacity of the challenge by Galileo to (1) the beliefs of contemporary leaders of the Christian church, (2) the ancient astronomical theories of Aristotle and Ptolemaic, and (3) the reigning common sense of simply looking up at the sky-Galileo was summoned before the Inquisition in 1633 AD. Galileo was forced to recant his position, and was banished to house arrest for the remainder of his life.

A little over a hundred years ago, a nurse recognized that when the surgeons were cleaner in the manner by which they performed surgery, the chances that a patient survived, greatly improved. The sentiment at the time was that a surgeon's gown, stained with the blood of his previous surgical cases, signified to his peers his surgical prowess and experience. Boldly challenging the common practice of wearing dirty surgical gowns into surgery-with common sense and undisputable data from her observations, the nurse spoke out. Recognizing the physical nature of the world that surrounded her, an audaciously brave nurse-not a physician, saved countless thousands of lives by implementing the simple concept of 'sterile technique', which remains the standard of medical care today.

As the twenty-first century unfolds, we can chose or not chose to recognize the physical world that surrounds us. We can chose to understand the mechanism behind century-old secrets as to how we chose a soulmate to spend the rest of our lives with, how ordinary individuals become geniuses, rich and famous, how our mind solves problems, what our subconscious really is and how we can communicate with it. There is a definable a mechanism that makes all of these wondrous things happen.

All men and women are morally created equal, but it is not clear whether all human brains are physically created equal. At least it may seem that there is a difference if you are taking a test in school and you happen to be seated next to a classmate who has a known 'photographic memory'. To your frustration, the classmate may appear to effortlessly answer the questions and turn his or her test in well before you do. What may really add to your frustration is knowing that you spent many hours studying for the test, while you know your classmate went out to a party the night before the test. Yet, after the tests are scored, this classmate ends up performing better on the test than you do.

How is it that some people are just plain smart? If all women and men are created equal, can any of us become a genius? Equality of our brains may be closer than we think. Becoming smart or becoming a genius may be a matter of better understanding our brains and rerouting its processing functions. Training our brain to do what we need or want our brain to do, may be analogous to training for a sporting event. The more you exercise in the proper fashion, the better one becomes to compete in a sport. More appropriately said in the context of this book, training the human brain might be analogous to exploring the features and range of the desktop computer. The more we understand how the brain is made, how it functions and therefore how to use the brain, the better our brain may perform. In a uniquely different direction, the more we understand the desktop computer, the better we will understand how the Human brain is constructed and the manner by which it functions.

This book describes the secrets of how the human brain functions, like a computer, so that you can understand how to explore and utilize your brain more effectively. It will cross you over to another zone of enlightenment, taking you to another dimension, so that.... *'You can get the most out of yours'*.

II. UNDERSTANDING COMPUTERS

1

WHAT IS A COMPUTER?

Computers today are represented by the sleek elegantly designed desktop and laptop devices most of us use at home or on the job. Not to long ago, the inner workings of a computer filled an entire room. Originally computers were devices meant to assist people with mathematical tasks. The word 'computer' is derived from the word 'compute', which means 'to solve a mathematical problem'.

The earliest way that man counted, except with his fingers, was by making scratches on bone, or creating symbols on flat surfaces covered with sand, or with objects such as shells, seeds, stones, or bones carried in a sack. The objects carried in a sack allowed primitive humans to perform simple addition and subtraction problems that were clear to the individual that owned the stones, shells or seeds and the individual the person was communicating or trading with. Later, primitive objects were replaced by valuable objects such as precious gems, and minted coins, which led to 'money' as a means of conducting commerce.

The earliest recorded computer was the abacus introduced in 2600 BC by the Chinese. An abacus is constructed of a wooden frame, with columns of beads or counters that freely slid along parallel wooden or metal rods inside the wooden frame. Typically there were six to twelve columns. Each column would typically have ten beads or counters, except for the last column on the right which might have only four beads, to facilitate splitting goods into quarter shares. Each column was designated with a different value of bead. The first column on the right consisting of ten beads typically would represent 'ones'. The second column of ten beads typically represent 'tens'. The third column of beads could represent 'hundreds', and so on. Experts with an abacus can be very proficient at solving addition and subtraction problems by manipulating the physical position of the beads. Abacuses were very commonly used by merchants for centuries, especially in the Eastern part of the world, for calculating quantity and sales of goods.

As time marched on, and technology progressed, the higher levels of math that developed required more sophisticated mathematical tools than what an abacus could provide. Multiplication and division problems were difficult to calculate on an abacus. Incorporating fractions of whole numbers would also be difficult on an abacus. Therefore, the answers to common mathematical calculations were performed by hand by mathematicians. Books comprised of tables of written answers to multiple common mathematical formulas provided a valuable, but slow reference source for architects, engineers, and commercial trading.

The slide rule was a tremendous step beyond manipulating beads on the abacus, using one's fingers to count, and flipping through pages of reference tables. The slide rule was comprised of two pieces of wood, metal or plastic, etched or printed with rows of markings and numbers, that slide in relation to each other. The slide rule offered a tedious, but effective means of performing multiplication and division. Log values could also be calculated. The slide rule worked on the principle that position of numbers on the two sliding pieces of the instrument allowed such calculations to be performed with great accuracy depending upon the skill of the user on making the position adjustments with the sliding bars.

Like the abacus, the slide rule allowed a proficient user the means to solve mathematical problems—but was limited because (1) positioning of the sliding bar was imprecise, (2) one had to remember place value of the number they were calculating (was the number in ones, tens, hundreds, etc.?), and (3) most important: only one mathematical problem could be solved at a time.

Early versions of modern computers were no more than simplistic calculators attached to a row of lights. These early electronic calculators could perform addition, subtraction, multiplication and division. The results were displayed as a row of lights either lit or unlit. Understanding what the lights meant required a working knowledge of the binary number system (based on ones and zeros) in order to decipher the result.

For example, the binary numbers that represent the digits one to nine:

BINARY NUMBERS
Binary Number=Numeric Number

0000=0	0101=5
0001=1	0110=6
0010=2	0111=7
0011=3	1000=8
0100=4	1001=9

The output device of the calculator evolved into an electronic light unit that was capable of displaying a row of numeric numbers, making it much easier for the average human user to make sense of the calculator's output. This was one of the first major steps towards making computers understandable or 'user friendly'. The early calculators offered rapid, accurate, and reliable answers to sophisticated mathematical calculations.

The breakthrough in the modern computer came with two outstanding innovations to the electronic calculator. Once was the advent of making calculators programmable—that is, the computer could be instructed to make a series of calculations before its task was considered completed. Second, answers to mathematical calculations could be stored and retrieved, providing the capability to utilize the answers in future calculations.

The most complex model of a computer that we have is the human brain. The human brain is capable of making mathematical calculations, of storing a variety of information, and taking learned information and conducting a variety of work. The human brain has a flare for creativity.

The desktop computer is quickly catching up to the human brain. The human brain can learn and assimilate vast quantities of facts. The human desktop computer is rapidly increasing its memory capacity and its processing speed. The 'human computer' is philosophically capable of determining (though not as in knowing) the difference between 'right' and 'wrong'—though this higher level of function is not always demonstrated by some members of society. Humans are commonly thought to use less than ten percent of their brain's capacity. As man-made computers expand their memory capacity and their processing power, they will be increasingly capable of storing larger and more complex programs. This expansion of memory and processing power, may edge the desktop computer closer to the philosophical capacity of the human user. Alarmingly, we are rapidly heading toward a time when computers may decide what is right and what is wrong...for us.

2

UNDERSTANDING THE NERVE CENTER OF THE DESKTOP COMPUTER: THE MICROPROCESSOR

Microchip/Microprocessor/CPU

A microchip is a wafer thin silicon chip. Inside this chip is an intricately designed piece of computer hardware. The microchip sports rows of spindly metal legs that this silicon nerve center uses to plug into a chip port on an electronic wiring board. Today's microchips contain hundreds to millions of transistors inside the silicon wafer body. The transistors in a microchip may be designed to act as a microprocessor, a memory unit, an electronic regulator switch, an amplifier, a detector, a sensor or other hardware capacity.

Microprocessors are microchips that process instruction codes and data to carry out and direct the electronic functions inside a computer. Computers may have a number of microprocessors, they may have hundreds of microchips. The microprocessor that acts as the central controller chip of the computer is known as the central processing unit (CPU). Highways of wire connections called 'buses' connect microchips with other microchips, storage devices and peripheral devices allowing information to flow inside the computer. The network of microprocessors, microchips and data buses comprise the nerve center of a computer.

Inside older computers, the CPU appeared as a long, thin silicon wafer with two rows of spindly silver legs. Gold offers one of the fastest transmission mediums to facilitate electron flow. The latest versions of the CPU appear more like a square, with numerous gold legs located around the entire perimeter of the silicon wafer.

Inside the flat wafer body of the CPU are millions of transistors. Transistors are the evolutionary product of the much older devices known as vacuum tubes.

Vacuum Tubes, Predecessor of the Transistor

Vacuum tubes are glass containers, generally with a plastic base (see Figure 1). Usually at least four metal components stood inside the confines of the exterior glass tube corresponding to at least five plugs at the base. One component was designated the cathode (a negatively charged electrode), one component was the heater, one component was known as the grid, and the final component was the anode (a positively charged electrode).

A vacuum was created, air removed from inside the tube to make it easier for electrons to travel from the cathode to the anode. A negative charge was placed on the cathode. The heater, placed in close proximity to the cathode, boiled electrons (negatively charged) off the top of the cathode. Electrons jumped through the vacuum in the glass tube from the cathode to the positively charged anode. The flow of electrons from the negative cathode to the positive anode represented an electric current.

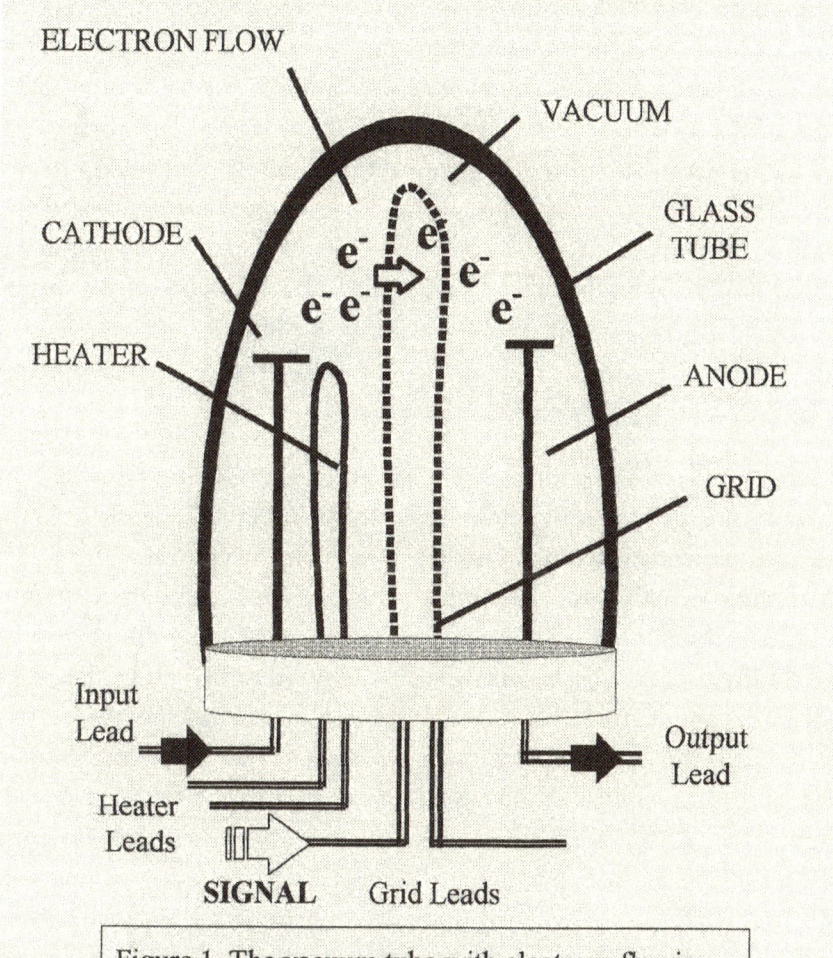

Figure 1. The vacuum tube with electrons flowing from the heated cathode to anode. A signal applied to the grid leads modulates the flow of electrons.

The grid, a device positioned between the anode and the cathode, would exhibit a negative charge. A strong negative charge on the grid would prevent electrons from jumping from the cathode to the anode by repelling the negatively charged electrons. A weak negative charge on the grid plate would allow the vigorous electrons to jump from the cathode and cross the vacuum to the anode. The strength of the negative charge on the grid plate dictated the number of electrons jumping from the cathode to the anode. Having the ability to control the electron flow traveling from the cathode to the anode, by changing the strength of the negative charge exerted by the grid plate, allowed for control of the output of the vacuum tube. A weak signal could be introduced to the grid, and directly control a much stronger output signal leaving vacuum tube through the anode lead.

The Power of Amplification

Radio waves traveling through the air are generally relatively weak in energy. In early electronic radio devices, incoming radio signals would be captured by an antenna and fed to a series of vacuum tubes. The weak radio signal would be applied to a grid lead of the first vacuum tube. The fluctuating incoming radio signal applied to the grid plate would cause the much stronger electron flow traveling from the cathode to the anode in the first vacuum tube to fluctuate. How the output current of the first vacuum tube fluctuated was therefore directly related to the incoming radio signal fed by the antenna to the grid lead. Now stronger, the generated signal would exit the first vacuum tube out through the anode and be fed to the grid plate of a second vacuum tube and be further amplified. An even stronger version of the original radio signal would then exit the second vacuum tube through its anode lead. The amplification process would be repeated using other vacuum tubes until the original radio signal was sufficiently strong enough to cause physical vibration in audio speakers. The sound produced by the vibration of the device's audio speakers is what a human would listen to with their ears.

Turning up the volume in a vacuum tube radio device, meant increasing the power to the cathode plate. The increased boiling of electrons off the cathodes resulted in a stronger modulated signal reaching the anode plate of the vacuum tubes, which resulted in a stronger signal being output by the vacuum tubes to the speaker, which would ultimately increase the vibration of the speakers, thereby increasing the sound emitted by the radio.

Volts and Amps

An imaginative way to understand electronic circuits is to think about the flow of water through pipes. Electron flow, otherwise referred to as 'current', could be thought of as flowing water. In an electric circuit, a power supply provides the force behind the flow of electrons. In the plumbing of a house, either a home pump provides pressure, or in an urban setting, the city provides water pressure either through pumps at the water treatment plant or by gravity if the water is stored above ground in a water tower. The flow of water from a higher to a lower elevation produces water pressure due to the gravitational pull exerted by the planet on the physical substance of the water.

In terms of electronic systems, the power behind the electrons is provided by batteries or the pressure is provided by a power plant. Power created by a power plant is derived from generators driven by the heat of the burning of coal, gasoline or oil, or the churning of turbines in a damn, or from the heat of nuclear fusion. The push behind the electron flow is termed voltage. The volume of electrons that pass through a conductive wire is termed current, which is measured in Amps or often in milliamps (one thousandth of an amp). In its technical terms, one amp is measured as the flow of 6 billion billion electrons traveling the length of one meter over the course of one second.

The importance of understanding concept of volts and amps can be painfully obvious, and may save you money. Recently, we took our home computer network apart for the purpose of removing the dust that had accumulated in the area surrounding the computers. One of my sons decided to reassemble his computer without asking for assistance. He did not pay attention to the power cord he attached to the scanner. When he turned his computer system on, he described the experience as a flash of light and a puff of smoke. He sought my help when he found that the scanner no longer functioned.

In reviewing the set up with him, we determined that the original transformer and power cord that had come standard with the scanner was rated as 120 volts and 25 milliamps. The transformer and power cord he had attached to the scanner, which looked very much like the original equipment, was rated at 120 volts and 35 milliamps. The power cord my son had attached to the now crippled scanner supplied the scanner with 10 more milliamps of current than the scanner was rated to operate with safely. In effect, my son had fried his scanner by running more current through the device than what the device was designed to handle; a simple mistake that cost $79.00 plus tax to replace the scanner.

The Development of the Transistor

Vacuum tubes worked well in their time, but were bulky in size and fragile due to the glass dome that surrounded the inner components of the tube, and were subject to wearing out. If the glass exterior of the tube cracked, the vacuum created inside the tube would be compromised and the vacuum tube would not work properly. Electronic devices made up of vacuum tubes were large pieces of equipment and easily damaged. In addition, vacuum tubes generated significant amounts of heat. When a vacuum tube was in service for long periods of time, excessive heat build up led to the glass exterior to crack or the electronic components inside malfunctioning due to misalignment or heat damage. The cathode lead of a vacuum tube often lost its capacity over time to boil off electrons. Modernization of the world of electronics came with the advent of the silicon transistor.

Silicon was found to offer attractive physical properties of electron conductivity. Positioning three silicon wafers together provided a means of creating a cathode, a grid and an anode together without having the bulky structure of a vacuum tube (see Figure 2). Electrons could pass from the first silicon wafer through the middle silicon wafer to the third silicon wafer without having to jump through the air. The middle silicon wafer could be energized to act in the same fashion as the grid plate of a vacuum tube.

Single transistors are easily recognized by the fact that they typically have three wire leads attached to a silicon structure or cap. One wire acts as a means for electrons to enter the transistor (the cathode, the resource), a second wire acted as an exit for electrons from the transistor (the anode, the drain) and a third wire acted as a means of regulating the flow of electrons (the grid, the signal lead). Transistors provided a replacement for the bulky glass vacuum tubes. The smaller size of the transistor meant that electronic equipment could be made smaller. The durability of the silicon meant that the electronic equipment could be made to withstand damage better, making electronic equipment comprised of transistors much more durable and reliable than equipment comprised of vacuum tubes. Heat generated by the flow or current was still a problem for the transistor, but the heat generated was somewhat less, and the physical properties of the silicon possibly tolerated heat better than vacuum tubes, offering longer lasting electronic equipment. Finally, the transistor did not wear out, like the cathode lead of a vacuum tube was subject to doing.

As the understanding of the properties of silicon advanced, it was learned that more than one transistor could be built into silicon wafer. Creating silicon wafers with more than one transistor led to the design of microchips. Today's microchips are designed with thousands, in some cases millions of transistors, built into a single chip.

Transistors offered the world the ability to create digital technology. Where the anode output of a vacuum tube generally fluctuated in response to the signal fed to the tube's grid lead, a transistor's anode lead could be made to produce a fluctuating output signal, or could be made to be either charged, or not charged, in response to the signal fed to the transistor's grid lead. Therefore, the output of a transistor could be either 'on' if charged, or 'off' if not charged. In electronics, the state of being 'on' is often represented by a 5 volt charge at the output, or 'off' is often represented by a 0.3 volt charge at the anode's output. This capacity by the transistor to be either 'on' or 'off' has lead to the use of computing devices that rely on binary mathematics. An 'on' output of a transistor can be considered a 'one' and the 'off' state at a transistor's output lead a 'zero'.

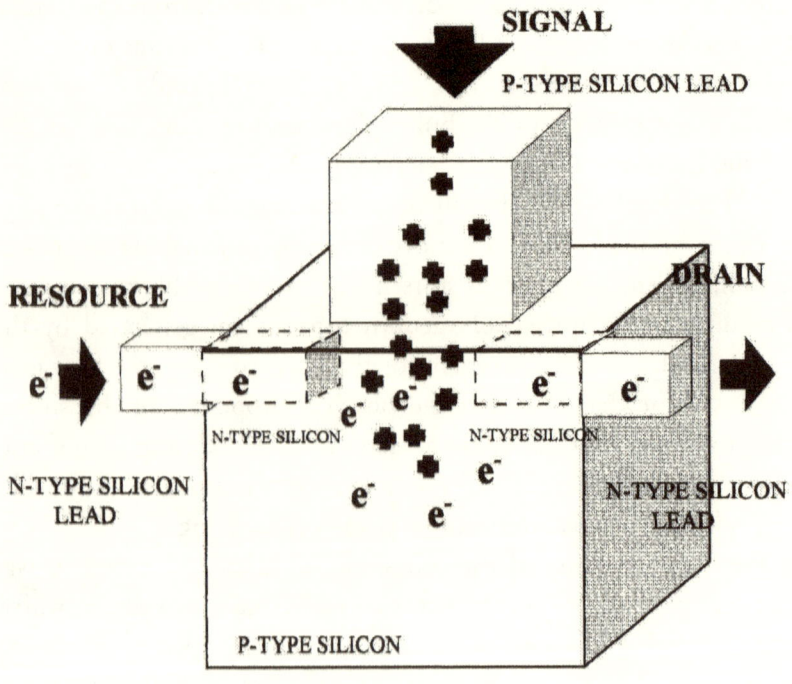

Figure 2. Sketch of a silicon dioxide transistor. The flow of electrons leaving through the drain is modulated by the signal input to the P-type silicon lead.

A Binary Computer Language

Early computer programmers communicated with the computer in a language that the computer understood, which was a series of 'on' and 'off' signals. The 'on' and 'off' states of a transistor could be taken to represent a 'one' and a 'zero' respectively. Each single on or off state of a transistor is considered a 'bit'. The computer could take finite strings of ones and/or zeros and perform functions on the basis of what a series of ones and/or zeros represented. A series of ones and/or zeros could represent a data string in some circumstances, while other series of ones and/or zeros may be specially coded to represent instruction commands for the computer to follow. This rudimentary language based on binary code that computer programmers and computers worked in has been referred to as machine code.

In a data series, a bit was considered a single on or off value. A byte was considered to be eight bits, or eight ones and/or zeros strung together. In a byte, from left to right, the value of the first one or zero represents a one or a zero. The value of the second one or zero represents a two or a zero. The value of the third one or zero represents a four or a zero. The fourth one or zero represents the value of eight or zero. The fifth one or zero represents the value of sixteen or zero. The sixth one or zero represents the value of thirty-two or zero. The seventh one or zero represents the value of sixty-four or zero. The eighth one or zero represents the value of one hundred twenty-eight.

In the binary number system, the number '00000001' is equal to 'one', where the number '00000101' equals to the value of four plus one, or five. See below:

BINARY CODE

(Example) Four Transistors=Binary Number=Base 10 Number

off off off off=0000=0   off on off on=0101=5

off off off on=0001=1   off on on off=0110=6

off off on off=0010=2   off on on on=0111=7

off off on on=0011=3   on off off off=1000=8

off on off off=0100=4   on off off on=1001=9

*Note: the first four bits of each byte are not represented (all would be 0).

The process of adding binary numbers together simply means that zero plus zero equals zero, zero plus one equals one, and one plus one equates to a one of the next higher value. Therefore, adding the binary number '00000101' (which equals five) to '00000100' (which equals four) results in the binary number '00001001' which is eight plus one or nine. See below for an additional example.

BINARY ADDITION

Four Transistors=Binary Number=Base 10 Number

First Number	off off on on	=	0011	=	3
Second Number	off on off off	=	0100	=	4
Add	+ -----------------	+	------	+	---
Total:	off on on on	=	0111	=	7

BINARY LANGUAGE

Terminology	Binary Number	Transistor output state
One bit	1 (or 0)	on (or off)
Two bit	11	on on
Four bit	1111	on on on on
Eight bit*	11111111	on on on on on on on on
Sixteen bit^	11111111 11111111	'on' repeated 16 times

* eight bits are comprise a byte, ^ sixteen bits comprise two bytes

Binary numbers can represent 'base ten numbers' (such as a programmer's data), or binary numbers can represent computer commands, instructions, or functions. Depending upon how the binary number is presented to the CPU, the CPU will either 'act' if the binary number is a command statement, or 'manipulate' or 'modify' the binary number if the binary number represents 'data'.

Placing binary numbers that represent command statements to a computer together in series is in effect a computer program. So computer programs, or otherwise known as 'computer software', is in effect, a series of command statements that often effect how data is used or modified.

A certain command statement will tell the computer's CPU to retrieve certain numbers from a specific location in a specific memory device such as the hard drive, the CD rom drive, the floppy disc drive, or from a source external to the computer such as possibly the Internet or an auxiliary hard drive device or disc drive. One command statement might instruct the computer to add two numbers together, another command statement might tell the CPU to subtract one number from another number. Another command statement might instruct the CPU to send the final result of the mathematical calculation to the printer to be printed on paper.

Writing a computer program in binary code is referred to as writing in machine language (the language of ones and zeros). The use of the binary language and writing assembly language programs is still the fundamental part of the modern day computer. We don't usually see this part of the computer because modern computers have higher levels of sophisticated, user-friendly program languages that the computer's user can interact with to get the computer to perform in a manner the user wants the computer to perform. The higher languages convert programmers' instructions and data into the computer's binary language, which the computer then either acts on or stores for future reference.

The CPU of the computer understands binary code. The CPU lives and breathes ones and zeros. Data and instructions are routed through the information highways called the data buses of the computer from a memory unit to the CPU and everywhere else by means of transporting finite strings of 'ones' and 'zeros' from one place to another. The speed of the computer is dependent, in part, on how large of a string of 'ones' and 'zeros' the computer can move at one time. The more bits that can be moved at one time, the faster information flows about the computer.

For example, if you were planning a large party and you wanted to move one hundred twenty-eight pop cans stored in your basement to the refrigerator located in the kitchen on the first floor of your home, if you could only move four cans at a time it would take you thirty-two trips. If you could load the pop cans into basket and move eight cans at a time, it would take you sixteen trips. If you could load thirty-two cans into a box, it would only take you four trips. If a friend helped you carry enough to move sixty-four cans at a time, it would only take you two trips from the basement to the refrigerator in the kitchen accomplish the job. In this example, you would save a significant amount of time, each time you reduce the number of trips required to complete the task.

Older eight bit CPUs could only move 8 bits of information at one time. A sixteen bit processor could work with sixteen bits or two bytes of information at a time. A thirty-two bit microprocessor chip is built with over thirty-two spindly legs extending from its silicon wafer body, and can process information streams comprised of thirty-two bits of information at one a time. The Pentium four microprocessor is capable of working with data streams that are sixty-four bits long. The greater the ability to move information in bulk, contributes to a faster operating speed of a microprocessor.

Humans enjoy the quick speed of the newer and more efficient processors, but humans have a tough time working with mundane things like ones and zeros. A video screen filled solely with ones and zeros just doesn't captivate our interest, and wouldn't sell too many computers on the consumer market. We relate better to letters of the alphabet from the written language we use every day. Further, we enjoy being stimulated by color and elaborate graphic designs, action, animation and even better, interaction.

Higher level languages were developed to make computers more user-friendly for the everyone to use, whether it be the stockbroker on Wall Street, the engineer seated in his office, the housewife who is acting as the secretary for the school PAT who has to type the notes from the last meeting, or the high school or college student surfing the Internet for references to write a term paper. Higher level computer languages turn the mundane 'ones' and 'zeros' swirling around the heart of the computer into brilliant colors, bright pictures, letters and audio sounds. So the 'ones' and 'zeros' of machine language interfaces with an intermediate language of the computer's operating system, which is written like shorthand, in computer code. This intermediate computer language is easier for computer programmers to write long programs with, but is still difficult for the average computer user to read or understand. The operating system of current computers, therefore, run computer programs written in even higher levels of user understandability and comprehension.

So for example: in utilizing a standard word processing program, the word 'example' is typed into the keyboard of the computer and then the enter button is struck. The word appears on the video screen before the user, but the computer functions that occur between the video screen and the CPU are quite detailed. The word 'example' is handed off from the word processing computer program to the intermediary computer program of the computer operating system which then hands the word 'example' to the machine language at the heart of the computer that breaks each letter of the word 'e-x-a-m-p-l-e' down into a computer recognizable string of ones and zeros. As a package, these strings of ones and zeros are directed by the CPU to the scratch pad memory for short term storage or long term memory in the Memory unit, if the user directs the computer to store the text on the hard drive in a memory file. The final data strings are also sent to the video monitor to be displayed so that the user who typed in the word 'example' can see and review the result of the typing. A graphics card connected to the video screen accepts the data string and interprets the meaning of the data, and is responsible for displaying the word 'example' on the video screen.

As technology continues to progress, the higher levels of computer programming actually start to share the intermediate computer language functions and the two actually are starting to share many functions together. Still, we produce computers and software programs that run at the fastest possible speed, CPUs need to be faster, computer programs need to be increasingly compatible.

3

THE CPU, MEMORY, &
DATA BUS

A computer is made of four essential components. These four main components include the Central Processing Unit (CPU), the memory unit(s), peripheral devices and the power supply. The CPU is the workhorse of the computer, it is the device that runs the computer. The memory device stores information that can be accessed by the CPU. The peripheral device or devices allow a human operator to communicate with the CPU and allow the CPU to communicate with the computer's user. The power supply converts household alternating current (AC) into direct current (DC) to energize the CPU, the memory and the peripheral devices so that these devices can interact with each other. The CPU is plugged into an electronic board called the mother board that allows direct connection to other devices plugged into the same board. The CPU and the computer's memory devices are connected together by strands of wire referred to as 'buses', which are much like a phone line. Information and instructions are passed between the computer components by means of a 'bus'.

The Central Processing Unit

A central processing unit (CPU) is the primary processor in a computer. The first recognized microprocessor to power a personal computer was Intel's 8088 chip. The 8088 was comprised of 29,000 transistors, operated at 4.7 megahertz, and could handle 8 bits of data at a time. Intel's 80486 microprocessor operated as fast as 133 megahertz and could handle 32 bits of data at a time internally. Intel's family of Pentium processors have continued to push the envelope of technology further. Intel's Pentium III is comprised of 9.5 million transistors, runs at speeds of up to 1,000 megahertz and can handle 64 bits of information at a time. The Pentium III (see Figure 3) in addition is equipped with two arithmetic logic units (ALUs) to handle the data. The advantage of having two ALUs allows the processor to work on two number sets interchangeably. There is an internal floating point number processor to manipulate floating-point numbers in an efficient manner. There are two types of memory cache units that allow the processor to hold data internally until needed: A pair of 16 KB L1 caches and a high speed 512KB L2 memory cache unit. Information moves 2 to 4 times faster by utilizing these internal memory devices rather than having to always send and retrieve data through the external bus to a memory storage device located outside the microprocessor chip.

Information enters the Pentium processor through the bus interface unit (BIU) and sends the information to the pair of L1 cache units. A fetch/decode unit pulls instructions in where three decoders, working in parallel, break down the instructions into 274 bit micro-operations, known as micro-ops. The fetch/decoder sends all micro-ops to the ReOrder Buffer (ROB) where the two ALUs handle all calculations involving integers. If a calculation involves a floating-point number, the ALU hands the job off to the floating-point math unit. A dispatch/execute unit checks the micro-ops in the ROB and executes them when they are ready to be processed. A retirement unit checks the micro-ops in the circular ReOrder buffer, and sends them to the store buffer when they have been processed, which then sends them to the BIU and then to RAM.

Intel's Pentium 4 processor is comprised of 42 million transistors. The Pentium 4 represents a redesign of the previous Pentium chip architecture. The Pentium 4 utilizes Intel's NetBurst microarchitecture to push data through the chip faster than previous models.

AMD brought to the market its Athon (or K7) processor, a 37 million transistor PC based processor, to challenge Intel's Pentium processors. The Athon is designed with a 128 KB L1 cache and a 256 KB L2 cache. It has three instruction decoders to translate instructions and feed them to the instruction control unit, which manages the execution and retirement of all micro-operations.

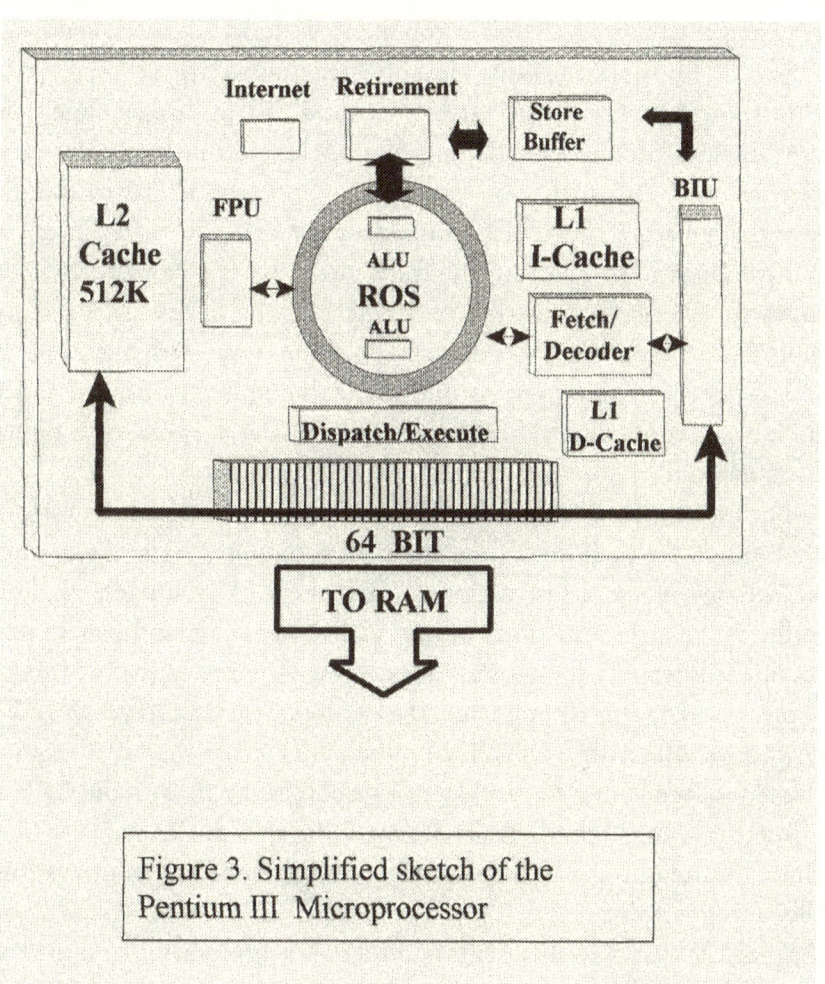

Figure 3. Simplified sketch of the Pentium III Microprocessor

The CPU, though the ringmaster or boss chip of the computer, is not the only microprocessor in a computer. The co-processors exist to serve the CPU and to perform duties to relieve the CPU of task burdens where possible. A co-processor microchip exists in devices such as the keyboard, some printers, graphics cards, and in soundcards.

The CPU is a device that seeks information, then initiates and completes tasks. A small internal clock provides regular impulses that drive CPU and the other electronic components to function at a specific rate of speed. This is not the same kind of clock that we are usually familiar with, though it is a clock that can be modified to display time. The internal clock produces regular impulses, ticking off intervals of time in microseconds (10 to the -6th power) or nanoseconds (10 to the -9th power) regulating the CPU's process of information. The faster the internal clock is able to run, the faster the CPU is able to perform its duties up to the construction parameters of the device. Mega Hertz refers to a million cycles per second. So a computer with a capacity of 400 megahertz is able to perform duties at a frequency or speed of 400 million a second. A gigahertz computer runs at a speed of a billion times a second.

The internal clock drives the CPU to function. In the ideal design of a computer, for each tick of the clock, the computer's processor would execute one or more instructions. Often the CPU waits idle ready to perform the next function, but is unable to initiate the next step because it is waiting for another device to send it a required instruction or piece of data. Pipelining refers the concept of trying to get the CPU to work as efficiently as possible by predicting what the CPU requires before it needs it, and have what ever resources supplied to the CPU in a timely fashion so that ultimately, the CPU spends the majority of its time executing programming functions and minimizing its time sitting idle, doing nothing.

The CPU looks to the memory devices for information and instructions. The CPU interfaces with various memory devices to accelerate function. In the Pentium chip itself, there are two high speed memory caches or devices. The Pentium III and Pentium 4 processors interface with RAM (random access memory) device by a 64 bit data bus. The RAM moves required software programs and information from other memory storage devices to its memory so that the program instructions and data can be quickly and easily accessed by the CPU. Memory devices continue to evolve improving speed of data access, amount of data access and durability. Common memory devices include the Random Access Memory (RAM), the hard drive, the CD-ROM drive, DVD drive, ZIP drive, JAZ drive, the floppy drive, and magnetic tape drives.

Memory devices are like a well-organized library. Instead of hard-covered books, in these libraries, the memory units store information in an electronic manner in the form of data files. These data files are meticulously arranged in the memory unit and can be recalled quickly to satisfy the needs of the CPU. When the CPU requests information, the memory device locates the proper data files from its memory banks, then sends the requested information back to the CPU through the information bus. The CPU then reads the information and acts upon the information.

The messages a CPU receives from the memory unit may be either numbers otherwise referred to as 'data' or an instruction or command. Instruction codes tell the CPU what to do next. Programs that computer programmers write are often long lists of instructions or commands that tell the CPU what tasks the programmer wants the computer to perform. This is often referred to as instruction code. As a collective, instruction codes are called computer programs and are stored in one of the computer's memory units. Therefore, as the internal clock ticks away, driving the CPU to process information, the CPU makes requests from the memory unit. A memory file, comprised of an instruction code, is sent to the CPU to direct it to the task to be performed next.

Tasks that the CPU is requested to perform are quite varied. The CPU may be asked to retrieve two numbers from the memory unit's memory files and add them together. So the CPU may request from the memory unit its next instruction. The memory unit may respond by sending an instruction code that tells the CPU to add two numbers such as <u>5</u> and <u>4</u>. The CPU will then request the first number. The memory unit will respond by retrieving the first number from its data files and sending this first number to the CPU; in this case a '5'. Once the CPU receives the '5', the CPU will then request the second number. The memory unit will then respond by sending the CPU the second number; in this case a '4'. Once the CPU receives the second number, the CPU will act as a calculator. The two numbers will be added together to get an answer of '9'. The CPU will then search for what to do with the number '9'. It will request the next instruction in the memory unit. The instruction code provided to the CPU will tell the CPU what to do with the answer derived from the addition. The CPU will either be instructed store this answer in the memory unit or incorporate this answer into another mathematical process, or display this answer on a video screen, or send this information to another computer by means of a inter-computer network link, or print this answer on a printer.

As the need for faster and faster CPUs pushes technology forward, CPUs have actually become more restricted in their function. The less the CPU has to do, the faster it can process instructions and read data sent to it from the Memory unit. An efficient CPU is one that can delegate tasks to other devices in the computer. Therefore, CPUs may have a separate 'logic' chip hardwired to it. The function of the logic chip is to perform mathematical tasks for the CPU. The Pentium III and Pentium 4 processors both have two Arithmetic Logic Units (ALUs) in the processor chip. If a CPU has a logic chip attached to it, then when the CPU gets an instruction from the memory unit to add two numbers together, the CPU can delegate that duty to the logic chip, thus freeing up the CPU to do other tasks while the logic chip performs the required mathematical function.

CPUs often have a scratch pad memory unit hardwired to that stores information for brief periods of time. A scratch pad memory is designed to hold small amounts of information for short periods of time, in this manner there is no delay in searching memory banks for information. Data that is needed to perform mathematical calculations can be retrieved quickly and easily from the scratch pad memory. Such scratch pad memory is accomplished by memory devices internal to the processor and by high speed RAM connected to the processor by a high speed bus interface unit.

Intel's Pentium III and the Pentium 4 processors have multiple internal memory devices. There are two L-1 caches with instruction code going to the L1 I-Cache and data going to the L-1 D-Cache. A second L-2 Cache comprised of 15.5 million transistors, capable of storing 512 kilobytes of data and code, provides high speed memory capacity. By having a scratch pad memory unit, the CPU is able to store and retrieve data, such as numbers in the scratch pad memory and use this data or numbers at is own pace. This frees up the CPU to run faster. Information is transferred back and forth between the processor's internal memory caches and RAM. The RAM unit transfers information back and forth from the other memory devices.

Data Bus

A data bus acts as an information highway, facilitating large amounts of data transfer from one point in the computer to another point. A data bus links the CPU to the RAM. Data ribbons (or buses) link the disc drives and hard drive to the motherboard. The motherboard is the circuit board that the CPU is plugged into. The mother board supplies the CPU with power and electronic circuitry, and may support the RAM, and expansion slots that accommodate a removable video card, sound card, network card and modem, to name a few.

4

PERIPHERAL DEVICES

Peripheral devices allow a computer to receive information and export information.

Input devices allow the computer to accept command signals or data input. Common input devices include a keyboard, a mouse, a light pen touched to a light sensitive screen, a microphone, a scanner, a modem, an optical lens, and an infrared sensor.

Output devices allow the computer to communicate with the computer's user or transfer information to a remote location or control a remote device. Output devices include video screens, printers, speakers and modems. A modem allows transfer of information to one or more computers in remote locations. Interface devices may allow a computer to direct the actions of machines such as a robot arm on a product assembly line.

Dual function input-output devices are considered to be the writeable CD device, floppy disc storage devices, magnetic tape memory devices, the modem, and the infrared port if the port is capable of both accepting infrared signals and outputting computer information through the same port such as used in some computer network configurations. A cradle or docking station may transfer information from a hand held mobile device to a desk top computer and vice versa thus acting as an input and output device.

The BIOS

When the computer is first turned on, the CPU must learn and recognize the internal and external devices the CPU is connected to and is expected to interact with. In addition, since the peripheral devices interact with the CPU, requesting services from the CPU, a system of prioritizing the requests for the CPU's time, amongst the various input and output devices, is required for the computer to function in an orderly fashion.

A system called the basic input and output system (BIOS) is the initial program that runs every time the computer is turned on. The BIOS identifies each component of the computer, and assigns an interrupt priority to the device. The CPU depends upon this priority designation in order for the flow of job requests to run smoothly. When the interrupt priority system fails, the CPU does not know which device to listen to or which task to attend to, and the result is the computer's CPU stops functioning, and the computer locks up.

5

PUTTING THE PERSONAL COMPUTER ALL TOGETHER

So you stare at this mysterious metal creature known as the desktop computer. Before you are the video screen, the keyboard, the mouse, the tower, the speakers, and the printer (see Figure 4).

You have a screwdriver in your hand. The tower of the computer has always mystified you. You figure it's either you or it. Somehow, you gotta know…what's inside that box. You haven't been compelled to rip open the box before now because it cost too much and you didn't want to permanently damage the mysteries that lay inside. But fortunately you just upgraded to the latest and greatest computer, a computer with an explosive Gigahertz processor, with tens of empty Gigabytes on the hard drive, a fancy DVD/rewritable CD drive and a screaming 256 Synched DRAM cache so that now your old computer, the one you are staring at, is headed on its way to computer heaven in some junk pile.

What's inside the box?…you ask.

Well, can the average human being make sense of all the little silicon chips, the circuit boards, cables and metal boxes? What do they all mean?…It all seems so sophisticated, complex, and overwhelming these days.

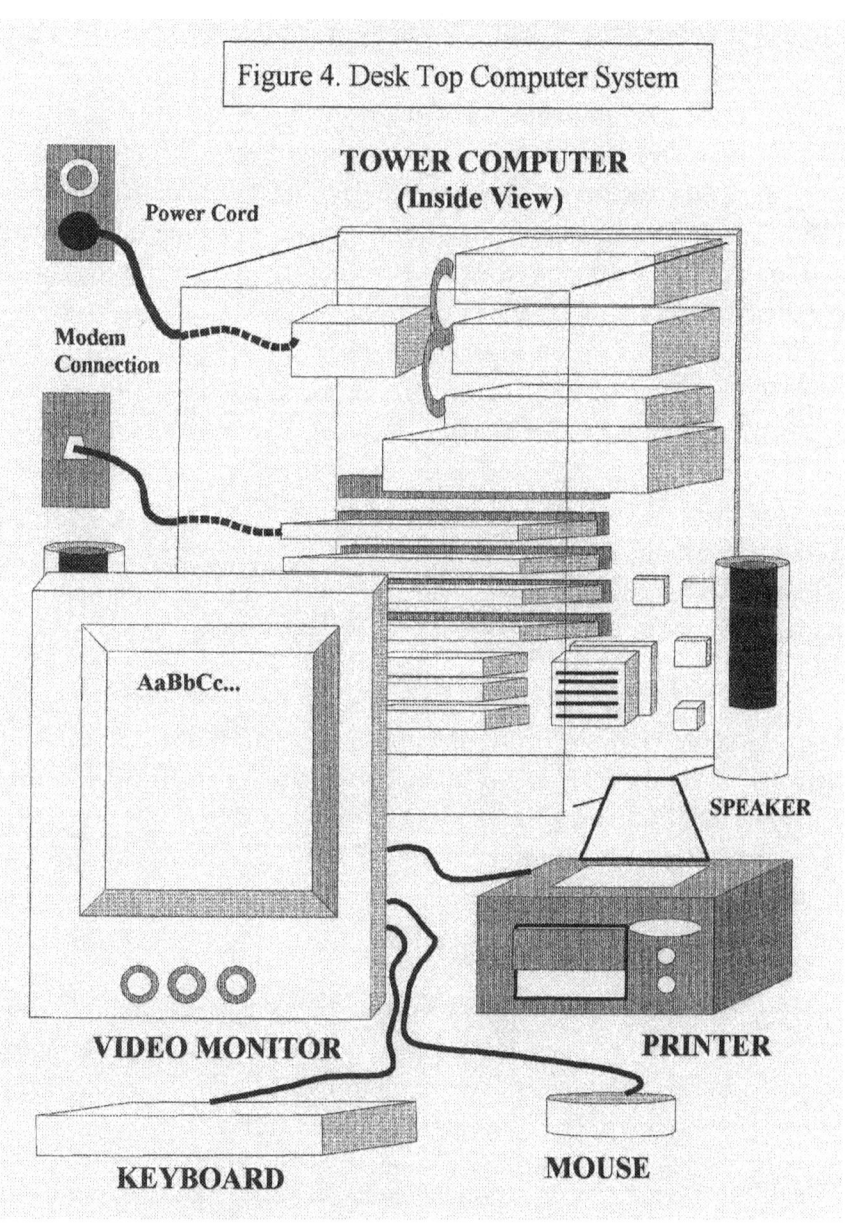

Figure 4. Desk Top Computer System

TOWER COMPUTER
(Inside View)

Power Cord

Modem
Connection

AaBbCc...

SPEAKER

VIDEO MONITOR

PRINTER

KEYBOARD

MOUSE

Actually, with the micronization of many of the computer components, there is a lot of air space inside the computer. The interior of the mysterious black box is actually relatively simple to figure out and understand.

The major components of the personal computer, often now referred to as a tower computer, because the guts of the computer now fit into a box that stands upright to facilitate the computer being placed on the floor next to a desk, rather than cluttering up the work area on top of a person's desk. Previously, desktop computers were constructed in a rather large box that rested on top of a desk. The computer's view screen sat on top of this rather obtrusive metal box. This original computer design, left little room for anything else to exist on top of an individual's desk. Placing the computer hardware in a tower, and putting it on the floor reclaimed the utility of an individual's desk, save for the video screen and keyboard that still require being on the desktop or fit into the design of the desk.

The tower computer is comprised of several discrete internal components that are easily recognized once the exterior metal shell is removed (see Figure 5):

Disc Drives: Hanging from the metal interior frame are the disc drives. Usually there are two or three disc drives. These disc drives may include a DVD drive, a rewritable CD drive, a CD ROM drive, a 3 inch floppy disc drive, and/or possibly a magnetic tape or ZIP drive. It wasn't too long ago that 5 1/4 floppy disc drives were standard in computers, but they are all but extinct now.

Hard Drive: The main memory device. This device appears as a box on the outside, but inside is comprised of one or more hard discs that spin at speeds of around 10,000 revolutions a minute. The hard drive holds the majority of the computer's software and data.

Power Supply: A power supply converts the 120v house current to a much lower voltage that the computer components can use. Cables from the power supply feed energy to the computer's interior electronic components. Much of the computer's components use 5 volts or less to effect the transfer signals and data. The mechanical devices such as the hard drive, disc drives, and cooling fans require higher voltages.

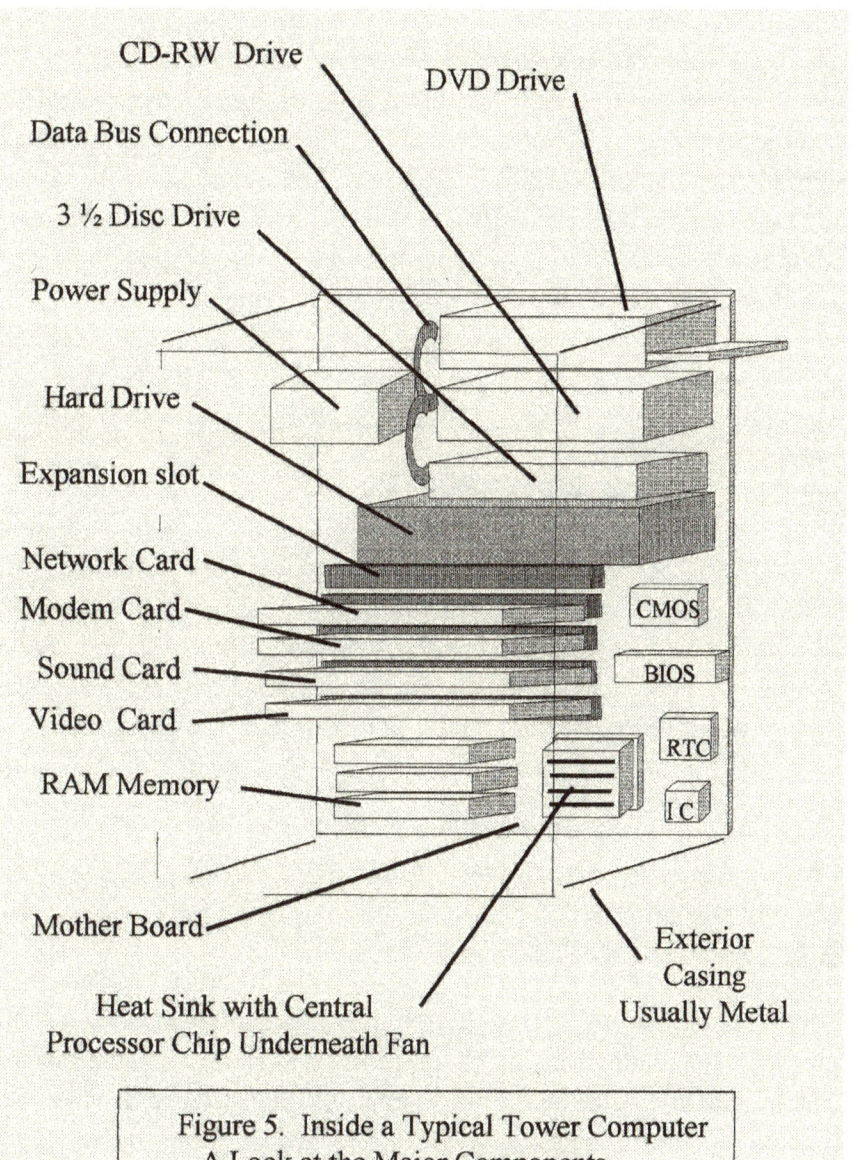

CD-RW Drive
DVD Drive
Data Bus Connection
3 ½ Disc Drive
Power Supply
Hard Drive
Expansion slot
Network Card
Modem Card
Sound Card
Video Card
RAM Memory
Mother Board
Heat Sink with Central
Processor Chip Underneath Fan
CMOS
BIOS
RTC
I C
Exterior
Casing
Usually Metal

Figure 5. Inside a Typical Tower Computer
A Look at the Major Components

Motherboard: There is a rather large integrated circuit board called the motherboard in the computer. The major components of the computer are either permanently soldered into this circuit board, plugged into sockets that are soldered to this circuit board, or connected by a data bus cable to this circuit board. Most personal computers today have only one circuit board. The need for other circuit boards has been reduced to cards, or miniature circuit boards that plug into expansion slots on the motherboard. The ability to remove these circuit cards allows for ease in repair of a damaged circuit, or the ease of upgrading a computer by taking out an old circuit card and simply plugging a new one in its slot. The central processing unit is generally plugged into a socket that is attached to the motherboard. Other components generally attached to the motherboard include CMOS, the BIOS, the interrupt controller, the real time clock and the RAM.

Central Processing Unit (CPU): The mastermind or brain of the computer. Older versions looked like long centipedes, with a long thin silicon wafer body supported by two rows of numerous silver legs. The latest CPUs are square in shape, with spindly gold legs extending from all four sides. At first the CPU may not be recognizable due to a fan or the fins of a heat sink that may be sitting on top of the CPU obscuring it from view. The CPU often employs its own fan to dissipate the heat generated by the electrons flowing through the millions of transistors inside its silicon wafer body.

CMOS: A memory chip attached to the motherboard that uses a battery to retain information regarding the computer's hardware configuration, even while the computer is turned off.

BIOS (*Basic Input/Output System*): Microchip(s) attached to the motherboard that is the heart of the computer. Stores the details of the hardware of the computer and serves as an intermediary between the software operating the computer and the hardware components comprising the computer.

Real Time Clock: A vibrating crystal that synchronizes and sets the pace for the work of the CPU and other computer components.

Interrupt Controller: a special microchip attached to the motherboard, interrupts the central processing unit when an immediate response from the CPU is required. Certain keystrokes, mouse commands, data or software events require the CPU to temporarily halt the activity it is engaged in, and divert its attention to the service identified by a higher order interrupt signal.

PCI expansion slot: Peripheral component interconnect slots are designed to facilitate plugging circuit cards into the motherboard to add or upgrade computer functions.

AGP expansion slot: accelerated graphics port gives 3D graphics cards fast access to the computer's main memory.

Video Card: plugs into a slot on the motherboard and translates image information into electrical signals needed to display the image on the video monitor.

Sound Card: plugs into a slot on the motherboard and contains circuitry to facilitate recording and playing multimedia sound. External jacks may be present for connecting speakers, headphones, a microphone or a CD player inputs.

Modem: a circuit card that plugs into the motherboard, that connects the computer to a phone line to facilitate accessing the Internet or sending/receiving faxes.

Network Card: this circuit card plugs into the motherboard to facilitate computer access directly to another computer or to a hub that connects the computer to a series of computers. Network cards facilitate the organized flow of information between computers. The link to other computers may be in the form of a twisted wire cable, fiber optics, phone lines, infrared light or radio signals.

RAM: random access memory is a collection of microchips that the CPU uses to store software and data while it uses the information. RAM acts as a memory storage device between the CPU and the hard drive and the other disc drives. When the power is shut off, the information stored in the RAM is lost.

Ribbon cables: or data bus, connect computer components such as memory devices to the motherboard and facilitate transfer of data between computer components.

Fan: a separate device from the smaller version that sits on top of the CPU. This fan draws in cool air to the interior of the computer to cool heat sensitive components. Some computers use cooling fins, that are simply slices of metal capable of dissipating heat as air passively flows across the fins.

See, it's all pretty straight forward. The external and internal architecture of the personal computer continues to evolve in size, shape, and resources as technology pushes forward. The exterior color changes, as fashion dictates.

6

THE LOGICAL WAY COMPUTERS OPERATE: HOW IMPORTANT IS *THE* INTERRUPT?

The central processing unit (CPU) is the processing workhorse of the computer. Most everything that occurs in the computer wants to utilize a portion of the CPU's processing power, and therefore, utilize some portion of time in the CPU's busy schedule. Since a CPU can only process one function at a time, some form of priority must be given to the functions that demand the CPU's time in order for computer functions to be tasked and completed in an orderly fashion. Therefore, a logic unit referred to as the Interrupt Controller (IC) (see Figure 6) is utilized by the CPU to determine which computer function has the highest priority at any given time, and therefore, which computer function is the next to be processed by the CPU. The interrupt controller uses a priority list either pre-programmed in the chip or accessible to it such as an interrupt table stored in RAM (read only memory).

Interrupts are a means of halting the current CPU process and interjecting a new process that has a higher priority than the process that the CPU is currently tasked to do. For example, if the CPU is working to process a computer program, but the CPU is requested to perform another task such as receive a fax on the telephone line, since the incoming fax file may be assigned a higher priority rating, the incoming fax may generate an interrupt command in the logic unit. The logic unit might respond to the interrupt command by halting work on the computer program and task the CPU to receive the fax file and store the fax file in an appropriate location in memory. Once the fax file has been properly stored and the interrupt command is removed, the logic unit may allow the computer program presently in the CPU, to continue to be processed.

The CPU recovers from an internal memory cache or the RAM the next instruction in the computer program that it had been working on and begins to execute the instruction.

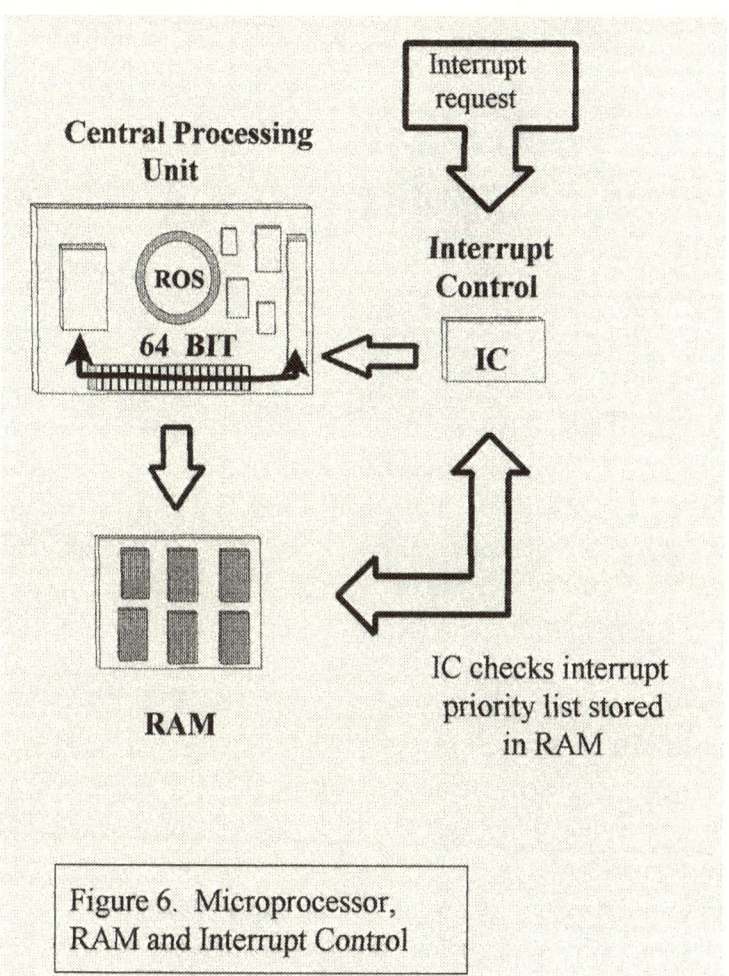

Figure 6. Microprocessor, RAM and Interrupt Control

7

THE SOFTWARE FRAMEWORK

At the core of the computer is the binary number system based on transistor technology. As previously discussed, the binary system is based on a series of 'ones' and 'zeros'. Ones and zeros are the only thing that a computer can understand, but very difficult for a human to work with. This rudimentary binary number system is a computer language referred to as machine language. Built on top of the machine language is a slightly more user friendly language called assembly language. Built on top of the assembly language is a more user friendly language referred to as an operating system. The language that operating systems are written in are more comprehensible to the average computer program writer, but is still very challenging for the average computer user to understand.

Several forms of operating systems have evolved. Microsoft developed DOS which stood for <u>D</u>isc <u>O</u>perating <u>S</u>ystem. IBM developed CP/M for control Program for Microcomputers. IBM also developed a graphic operating system called OS/2. Other operating systems include GEOS and DR DOS.

Much of the computer world is most familiar with Microsoft's DOS operating system. DOS provided commands that a computer user could utilize to run, copy, save, and otherwise manipulate computer programs and data that were run on a computer.

In the not too distant past, in order to run a computer program, the operator needed to turn the computer on, insert a floppy disc and load the DOS system. Once DOS was loaded, the disc was removed and then the program the computer operator wished to run could be loaded and run. Because memory space was so limited in the earlier computers, every time the computer was turned 'on', DOS, the computer's operating system, had to be reloaded. Every time the computer was turned 'off' the computer's memory was wiped clean. Only the information that was purposely stored on a floppy disc existed beyond the turning 'off' of the computer.

Microsoft developed Windows, which evolved from Microsoft's original Disc Operating System. Windows added a much greater versatility, a whole new dimension of user friendly operations to the computer operator, and allowed the opportunity to run more than one function at a time. Most currently marketed computers run Windows as the main operating system.

As most computer users are aware, there exists a flood of software programs on the market for purchase that exist in all forms of categories that can be bought and loaded on the average computer. Today's programs tend to be very interactive and user friendly. Despite all of the colorful graphics, bells and whistles,...all computer programs still boil down to the basic use of 'ones' and 'zeros' of the binary computer language in order to accomplish the tasks they perform.

An elementary computer program might appear as:

Line 001   A=4

Line 002   B=5

Line 003   C=A+B

Line 004   Print C

Which if you instructed a computer to RUN (or execute) this program, the result would be that the printer would PRINT the results of the addition of A (=4) to B (=5) which would be the number '9'.

A slightly more versatile computer program might appear as:

Line 001	Input A
Line 002	Input B
Line 003	Input X
Line 004	If X=1, then C=A+B, then Go to 006
Line 005	C=A-B
Line 006	Print C

In this scenario, if you requested that a computer RUN this program, the video screen would first prompt you to input a value for A, and then a value for B. After having done so, the computer would then prompt you to input a value for X. If you typed in '1', the program would add A to B and print the sum. If you typed in any number other than '1' when prompted to enter a value for X, the computer would subtract B from A and print the result.

Twenty-six years ago, the above example would have been a valuable learning tool. By today's standard, with advancement in user-friendly features of computer software, only the computer programmers worry about lines of computer code. The rest of us just enjoy the special effects and dynamic interactive features that working with higher order computer programs have to offer.

8

BOOTING UP YOUR COMPUTER

An operating system is a complex computer system that establishes rules on how a computer will run. Without an operating system, a computer cannot run except in maybe the most rudimentary form. When a computer is first turned on, it is analogous to a newborn baby. The computer initially has only enough computer code to send it looking through its memory for the BIOS program and for the files of the operating system by which the computer will run. This search for an operating system and the loading of the operating system is termed booting up.

Many of us are unaware of the current booting up process computers go through before we can use them. As mentioned previously, in the not to distant past, the meaning of booting up a computer was painfully clear to the average computer operator. In order to run a computer program, the operator needed to turn the computer on, insert a floppy disc, and load the computer's operating system before the computer would run. Once the Disc Operating System (DOS) was loaded, the disc was removed and then the program the computer operator wished to run could be loaded.

Presently, we are only aware of the booting up process when we get irritated by the newer computers as they either seem to (1) take an exorbitant amount of time to become operational once we turn them on, or (2) when the computer crashes and we find ourselves scrambling to locate the emergency recovery disc in hopes of rebooting the computer so that it will start again.

But booting up the computer remains one of the vital internal functions of the computer. Without an operating system to manage the flow of information, and to run the computer itself, and run the computer programs loaded on the computer, the computer would not operate efficiently or effectively. Without having the basic input/output system check the devices connected to the computer both internal and external, and checking interrupt priorities assigned to each device, the CPU would not be able to conduct a smooth flow of operation.

Once the computer is turned on, electrons stream through the vast maze of transistors, resistors, capacitors, diodes and other hardware in the computer. The microchips are energized. The input and output devices come alive.

If all one has on his or her computer is the operating system, they quickly learn that their computer can be pretty boring box that takes up a lot of precious room on the top of or next to their desk. We teach our computers how to serve us to our delight by loading software programs that we have interest in. If we use our computer to assist us with duties related to our profession, then professional software for writing or design tools are loaded on the computer. If the computer is used only to access the Internet, then the computer may have only Internet launching and surfing software. If one's computer is only used for entertainment purposes, then the computer's memory space may be filled with software games, or audio and/or video recording, editing and playing software.

The idea is, that a computer is otherwise a blank box until it is instructed as to what we desire to teach it through the software programs that we load via CD's or floppy discs or downloaded off the Internet or local computer net into our computer.

To take this thought one step forward, computers are constructed and sold all over the world. The computers bought and sold on the world market all run by the same binary number principle. The higher-level computer languages that are loaded on these computers located in different parts of the world are written utilizing different spoken and written human languages, in order to accommodate the different language needs of people across the world. You can virtually take any computer and load on to it a computer program written for a different language, other than let's say the English language, and it will display back to the computer's user output text in this different language as long as the software is written properly for the computer in question and properly written to display the information in the desired language.

Therefore, booting up today's computer, to make the computer a useful item planted on our desktop, includes loading an operating system, loading software and data, and having the output of these software programs displayed in a human language that we can easily understand, whatever that language may be, as long as there exists software written in the language.

9

NETWORKING YOUR COMPUTER: AIR, FOOD,... THE NETWORKS UP AND RUNNING

The sharing of information from one computer to another has recently evolved into one of the most powerful assets that a computer has to offer.

There are essentially three different ways to network a computer. One way is to connect the computer directly to a second computer. Directly connecting one computer to another requires both computers to have a network card, networking software and a connecting cable. The network card sees to it that there is a free, but orderly, flow of information between the two computers. The computer's network card first checks to see if the cable line is clear before it sends a message or data string to the other computer. If there was no order, both computers might be attempting to talk at the same time and no meaningful information would be passed from one computer to the other. With network cards and network software, one computer is placed in a listening or receiving mode, while the other computer is in the talking or sending mode. This method of connecting the computers directly to each other and each computer exhibiting the same amount of control is called a peer-to-peer network.

The peer-to-peer network concept can be expanded by connecting several computers that possess network cards and network software to a device called a hub. The hub may have five or ten or more computers attached to it. The hub acts to accept information that is sent to it from a computer and reroutes the information to another computer that is connected to it.

As technology has evolved, networked computers do not have to be physically connected. Infrared technology has provided the capacity to connect computers together within the confines of an office without having to run cables between the computers. Further, wireless phone technology has provided the means of connecting mobile computers, even hand held computers, to other computers.

A second methodology utilized to network computers is to construct a client-server computer architecture. One or more of the computers in the network is designated to act as a server computer. The server computer is loaded with the main operational software and stores the database. Other computers connect to the server through physical cable lines through one or more hubs, or through infrared or wireless communications. The client computers possess software sophisticated enough to access the main server program stored on the server computer. The concept of a client-server network is that the database is kept in only one place and client computers all work off of one database. If a single client computer suffers a computer problem, the remainder of the network continues to function properly for the remaining clients. If the server computer suffers a problem then the entire network shuts down. The network may be quickly restored by designating an alternative computer as the server and loading the alternative computer with information backed up to a floppy disc, CD device or magnetic tape prior to the occurrence of the malfunction in the original server computer.

A third way to network one's computer is through a modem that connects the computer to the Internet by means of a telephone line or cable line or wireless telephone. One's computer must possess software to connect their computer to host computer that will, in turn, make a connection to the World Wide Web. Many computer users have become very comfortable in logging onto the Internet, exploring what the Internet has to offer, and downloading information that one might find useful. Some computer operators maintain a continuous connection between their computer and the Internet.

Networking has become almost an indispensable, and certainly a very powerful function of the personal and corporate computer.

III. HUMANS AS COMPUTERS

B
THE PERFECT IDENTITY SWITCH

Nevin Thomas showed up on the twenty-second floor of the city medical center. He wrapped his bare knuckles on a sterile white wooden door. He coughed. A pain stabbed him, originating from deep inside his chest. He swallowed hard. The coughing was becoming more frequent. The sour taste of blood often nagged at the back of his throat. The shortness of breath was becoming noticeable and was limiting how far he could walk these days. Two chest surgeries and the best chemotherapy that money could buy had only slowed things down, hadn't cured him.

Dr. Christopher Stephens creaked opened the door. Through his gray horn rimmed glasses, the doctor peered at the caller. Recognizing his patient, the doctor waved Mr. Thomas into the clinic.

As Dr. Stephens secured the door behind Mr. Thomas, he asked, "Do you have the money?"

Tucked under his right arm were two curled up brown paper bags. With confidence Nevin responded, "Yes, I do." Nevin gave the doctor the first plain paper bag. He added, "I'll send you the balance in a couple of weeks."

The doctor nodded. He then stretched out his right hand ushering Nevin Thomas in the direction of a back room.

Nevin, his dark brown eyes open wide, his palms moist with sweat, anxiously asked, "Is David ready?"

Dr. Stephens curtly responded, "He's ready."

At fifty years of age, Nevin had been diagnosed with an incurable cancer. He had battled the slowly growing tumors in his chest for 18 years. His life was now in jeopardy. His time was ever so short. He had no medical options, save one, the option that Dr. Stephens offered eighteen years earlier.

Where at one time the procedure had been only a dream of Dr. Stephens, it was now almost a reality,…with funding from Nevin Thomas, Dr. Stephens and his colleagues had been able to crack the brain's basic operating system software codes. Dr. Stephen found he could communicate directly with the software program in a person's brain, much like one might be able to communicate with the operating system of a desktop computer.

Dr. Stephens ushered Nevin to a chair. Nevin Thomas casually dropped the second paper bag he carried on a nearby empty counter top. Nevin then dropped into the soft cushions of the rather large lazy-boy chair. Nevin pushed backward, and the footrest automatically rose, placing him in a reclining position.

Dr. Stephens plugged a computer cord into the plastic computer port that had been previously surgically implanted into the back of Nevin Thomas's skull. David Thomas, Nevin Thomas's supposed nephew, lay unconscious in the second chair. A computer cord was plugged into the back of David's scalp.

Nevin asked, "Can he hear anything?"

Dr. Stephens replied, "Doubtful,…But if he does,…it wouldn't mean anything to him…He's well sedated."

Nevin asked, "Has his mind been wiped clean?"

Dr. Stephens worked the dials on his computer console. The bio-computer genius responded, "His upper cortical function has been cleared."

"How is his health?"

"Ran him through a physical earlier today. He is a perfect specimen."

"Will he get the cancer that I have?"

The physician speculated, "Being he is your genetic twin, and clone, maybe in about thirty years…but at the rate medicine is advancing, they may have a cure for your kind of cancer by then." The doctor added,

"And if he,…I mean you don't smoke,…you may not face the same cancer you have now."

Nevin Thomas relaxed in the plush chair. Dr. Stephens meticulously worked the dials on his computer console. Dr. Stephens was so excited, his hands were jittery, his mouth was dry. Dr. Jim Cleo and an assistant George Tress walked into the room. In an excited voice Dr.

Stephens exclaimed, "It's finally happening! We're gonna make history."

Dr. Cleo calmly asked, "Any hitches?"

Dr. Stephens replied, "Not a one…The download procedure appears to be working like clock-work. Both men's brains are responding to all of our software commands just as predicted."

The three men dressed in white coats looked on as the two patients lay silently in the chairs before them.

The lab's computer worked feverishly to extract Nevin Thomas's brain files from his cerebral cortex, briefly storing them, then uploading Nevin Thomas's brain files into the genetically identical brain in David Thomas's head.

Once the upload was complete, Dr. Stephens removed the connection to David Thomas's brain. While Nevin Thomas's brain was still connected to the lab's computer, Dr. Stephen reloaded a brief software program into Nevin Thomas's otherwise now empty brain.

Dr. Stephens, Dr. Cleo and George Tress looked up from the lab's main computer screen. The work was done. All three men felt their hearts pounding rapidly in their chests as they anxiously waited to see what results they would have. A sixty eight year old man sat limp in the first chair. A vibrant 18 year old lay asleep in the second chair.

Dr. Stephens shuffled over to Nevin Thomas and disconnected the computer interface plugged into the back of the man's head. He then used a scalpel and with George Tress's assistance, carefully removed the computer interface plugs from both men's scalps. Using sterile hemostats the wounds were quickly sewed closed with resorbable stitches.

An hour later, David woke up flashing his young brilliant eyes wide open. After assuring his doctors he felt perfectly fine and thanking them for their efforts, David went to exit the surgical suite. Just before leaving, David Thomas asked, "You're gonna make it look like an accident, right?"

Dr. Stephens nodded, confidently remarking, "It certainly should."

David put his right hand to the back of his head. His scalp was extremely tender where the plug have been removed. David remarked, "The back of my head is awfully painful Doc."

Dr. Stephens replied, "Take two aspirin when you get home, and put a cold compress on it." He added, "I don't want to give you anything that might be stronger, and centrally acting. I don't want to interfere with your brain's neurotransmitters at this early stage."

Gruffly David grunted, "Sure."

Dr. Stephen stated, "See me in a week for follow up…At my house,…Not here."

"Got ya Doc," was the young man's quip reply. With that, David Thomas exited the surgical suite.

Fifteen minutes later, with some considerable prompting by Dr. Stephens, Nevin Thomas was aroused. He rose up out of the plush chair. Without saying a word, the shell of the man left the surgical suite.

Nevin walked down an empty hallway. He stepped onto a crowded elevator. The doors closed with a swish and the elevator car descended. A blank stare filled the aging man's eyes. On the ground floor the two elevator doors opened. Nevin was the fifth person off the elevator car. He felt like a steer being herded into the building's main lobby. He walked with the flow of bodies, most headed out the front door of the building.

Nevin passed through a spinning door. The pupils in his eyes stung from the intense brightness of a late afternoon sun. A crowd of strangers walked by on the bustling sidewalk. He melted into the midst of the crowded sidewalk. He slipped through the river of people, bumping a few, and feeling almost as if he were choking, claustrophobic from the congestion. Struggling through the crowd, he reached the other side of the stream of pedestrians passing by on the busy sidewalk.

The tall dark haired figure stood watching the traffic on the street. After diligently standing his ground for a few moments a big, silver bus, decorated with wide bright red stripes on the face and sides of the vehicle, came charging down the busy road.

As the bus barreled down the far right hand lane, Nevin felt a nudge from behind. The remnants of memory in his mind instructed the shell of a man to robotically walk out into the street. The nudge from behind initiated his forward momentum. With a blank expression masking the granite stone features of his face, unflinchingly, Nevin stepped off the sidewalk curb and stepped out into the path of oncoming traffic. The towering structure of a city bus, traveling at thirty miles an hour, steamed toward him. The bus driver had no way of stopping in time. The fifteen-ton bus struck Nevin head-on. The man's body was crushed by the initial impact, then run over by the deep treads of the huge black rubber wheels of the bus. Every neuron in Nevin's body suddenly exploded in a wild flurry of sensation, but Nevin never felt a thing.

Down the block, hidden in the river of pedestrians, a twenty-four year old with rich, dark mahogany brown hair, licorice shaded sunglasses bridging her soft, petite nose, casually, but hurriedly walked through the crowd. Her divinely sculptured hips temptatously swayed back and forth. Behind her a large city bus screeched, its brakes quickly overheating as the rubber tires fitfully tried to come to an immediate halt. The wheels locked up. The rubber treads burned the pavement as the bus skidded to a stop. As the eerie sound resonated through the street, the half of the crowd streaming away from the scene of the fatal accident stopped in their tracts to try to catch a glimpse of the tragedy. The other half of the pedestrians walking toward the accident scene continued on their way, hoping to get a better view of the gory details of the tragic scene. The brunette stopped momentarily and glanced over her shoulder. She sensed by the response of the people closest to the scene, that the man she had bumped had indeed met the end of his life.

She smiled, turned, and proceeded down the street heading away from the accident scene.

An unobtrusive earpiece was cradled in her left ear. The encrypted cell phone in her pocket picked up an untraceable signal. The message was simple, the voice muttered, "I'm out." Then the line went silent. The phone connection was disconnected. Amongst the thousands of calls made every minute in the city, they were hoping the encrypted message would not be detected by the authorities.

Casually, with her left hand, the brunette plucked the earpiece from her ear and shoved the earpiece into her left pocket. She then stuck her right hand deep into a pocket hidden in her sweater. The long, thin, spindly fingers of her right hand wrapped around a black palm-size box with a single button mounted on its face. Her actions concealed by the bulky clothing she wore, she pressed the button. A second later, behind her, the plate glass windows of the twenty-second floor of the medical office building were suddenly blown out, as the repercussions of a large blast resonated through the air.

A man shouted, "Hey look at that!"

Overhead, smoke spewed out from the flames that burst forth engulfing the twenty-second floor.

A shrill wail rose from the panicked crowd that filled the walkways and streets.

The brunette nonchalantly continued on her way, casually melting into the crowded sidewalk, swaying her dangerously curved hips from side to side as she strutted away, putting distance between herself and the crime scene.

10

ANATOMY OF THE HUMAN BRAIN & NERVOUS SYSTEM

Neurology

Neurology is the study of the brain, the different components of the nervous system, the diseases and injuries that occur to the brain and nervous system, and how to medically treat the brain when disease or injury occurs.

The Brain in General

In simple terms, the human brain resembles the shape of a football helmet. It has a dome shaped top portion. Two lobes that extend down on either side. In the center of the brain is the brainstem that acts as the information highway between the upper portions of the brain called the cerebral cortex and the remainder of the body.

Descriptive terms identifying positions of the body are important in being able to orient oneself to the location of parts of the brain in relationship to other parts (see Figure 7).

Anterior or Frontal: refers to objects located in the front portion of the body.
Posterior: refers to objects located in the back of the body.
Medial: refers to objects located close to the midline of the body.

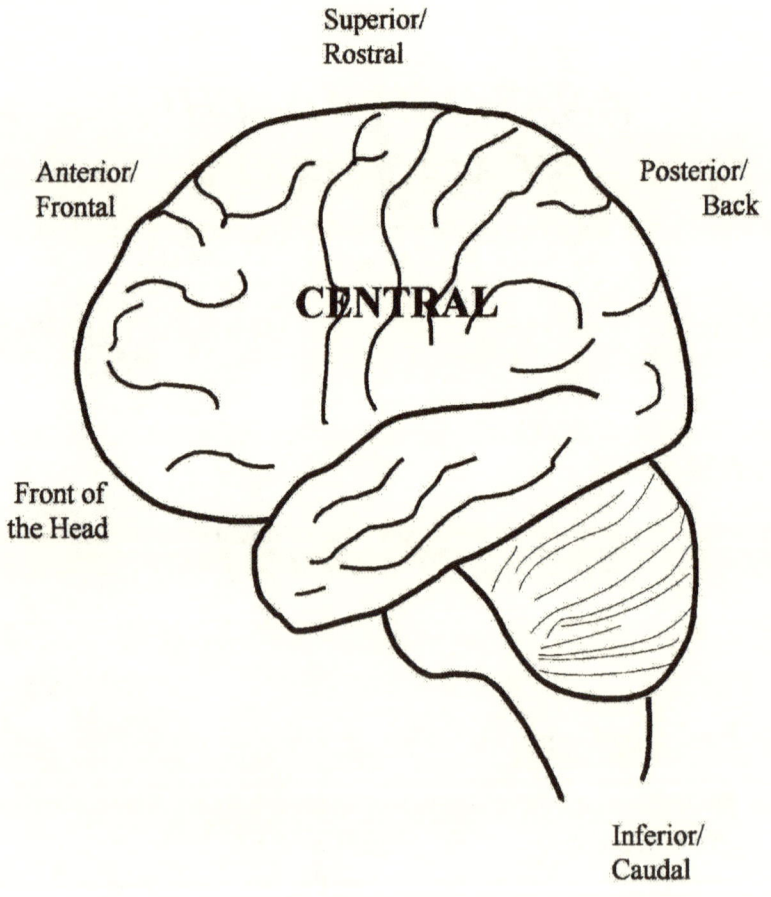

Figure 7. The Human Brain (left side view).

Lateral: away from the midline, toward the outside of the body.
Superior: refers to an object above or on top.
Inferior: refers to an object located below or on the bottom.
Central: refers to inside the brain and cranial vault.
Peripheral: refers to objects outside the brain.

Rostral: refers to an object located toward the top.
Caudal: refers to an object located near the bottom.

The upper portion of the brain is termed the cortex. 'Cerebral' is an alternative term that refers to the brain, so cerebral cortex refers to the upper portion of the brain. The cerebral cortex is considered the controlling portion of the brain, where the higher level or the more sophisticated brain functions occurs. Our sense of self-worth is thought to be somehow attached to this cerebral cortex.

The lower central portion of the brain is the brain stem. The upper portion of the brainstem is called the midbrain. The midbrain is comprised of a number of nerve nuclei. Nerve nuclei are where a number of nerve endings related to a common body function congregate together. The function of the nuclei range from accepting data from the special senses such as taste receptors on the tongue, olfactory sensors in the nose, and the cochlea in the ears; to routing information from the spinal cord to the cerebral cortex and then back down to the body.

In the lower rear of the brain resides the cerebellum. The cerebellum is a rudimentary or primitive portion of the brain. Sliced in half, the cerebellum resembles a Christmas tree turned on its side. The cerebellum is best known for controlling body position and balance.

Residing on top of the brainstem, the cerebral cortex is divided into several significant parts. The front of the cerebral cortex is termed the frontal lobes. Behind the frontal lobes, comprising the mid portion of the cerebral cortex are the parietal lobes. The rear of the brain is made up of the occipital lobes. The lower midbrain region on both sides has a tube-like projection designated the temporal lobe. The brain is divided into a right and a left side called hemispheres. The brain appears approximately to be physically equal on both sides. Therefore there are two frontal lobes, a right side and a left side. There are two parietal lobes, two occipital lobes, and two temporal lobes.

Right-handed people are believed to have the left side of the brain as their dominant side. This notion of the left brain dominance is dictated by the fact that in the brain stem the nerve fibers that originate in the left side of the brain cross over to the right side of the body and travel down the spinal cord to eventually control the muscle groups comprising the right arm and right leg. But it is actually not right hand or left hand dominance that determines which is a person's dominant hemisphere, it is which side of the brain controls speech that determines the dominant side of the brain in the eyes of a neurologist.

The frontal lobes (see Figure 8) are considered to be where personality and emotional centers are located. Inferior (lower) portion of the frontal lobe in the dominant side of the brain generates speech. The parietal lobes control movement and sensation in the body's arms, legs and trunk. The parietal lobe in the dominant hemisphere of the brain also interpret and creates speech. The rear occipital lobes represent the vision center, the area where what our eyes see, is interpreted as recognizable pictures in order that the brain may understand what exists in our direct line of sight. The temporal lobes facilitate long-term memory and interpretation of music.

The brain stem is comprised of a numerous nerve pathways. The function of the brainstem is to provide a pathway for the flow of information routed into the brain from the body as a whole to reach the important portions of the upper brain. In addition to processing incoming information, the brainstem facilitates the transmission of outgoing information and commands generated by the brain.

Buried deep inside the brain, located above of the brainstem, are a number of complicated structures are believed to route information up from the spinal cord to the cerebral cortex and from the cerebral cortex down to the brainstem, the cerebellum and to the remainder of the body. Near the top of the brain stem is the medulla oblongata (MG), above the MG is the Pons. Located just behind the Pons is the cerebellum. Just forward, and separate from the Pons is the pituitary gland. Above the Pons is located the thalamus (see Figure 9). Above the thalamus is the corpus callosum. The corpus callosum is a rather large curved structure at the center of the brain that stretches from the front of the brain to the posterior of the brain. Spanning the thalamus on either side is the caudate nucleus. Between the thalamus and the lentiform nucleus is the internal capsule. Above and surrounding and covering the corpus callosum is the cerebral cortex. The bony structure that surrounds the cerebral cortex is the skull.

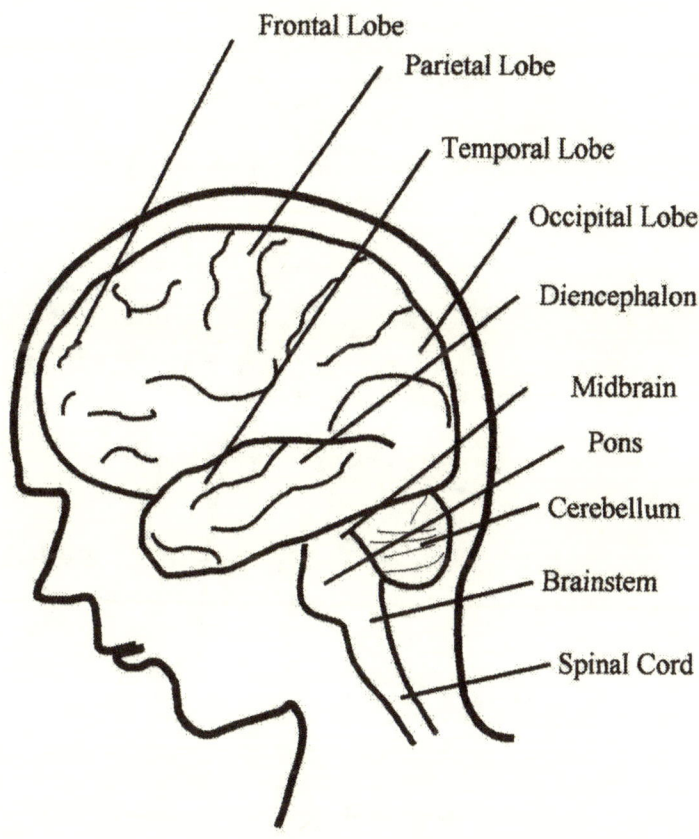

Frontal Lobe
Parietal Lobe
Temporal Lobe
Occipital Lobe
Diencephalon
Midbrain
Pons
Cerebellum
Brainstem
Spinal Cord

Figure 8. Major Divisions of the Human Brain (left side view).

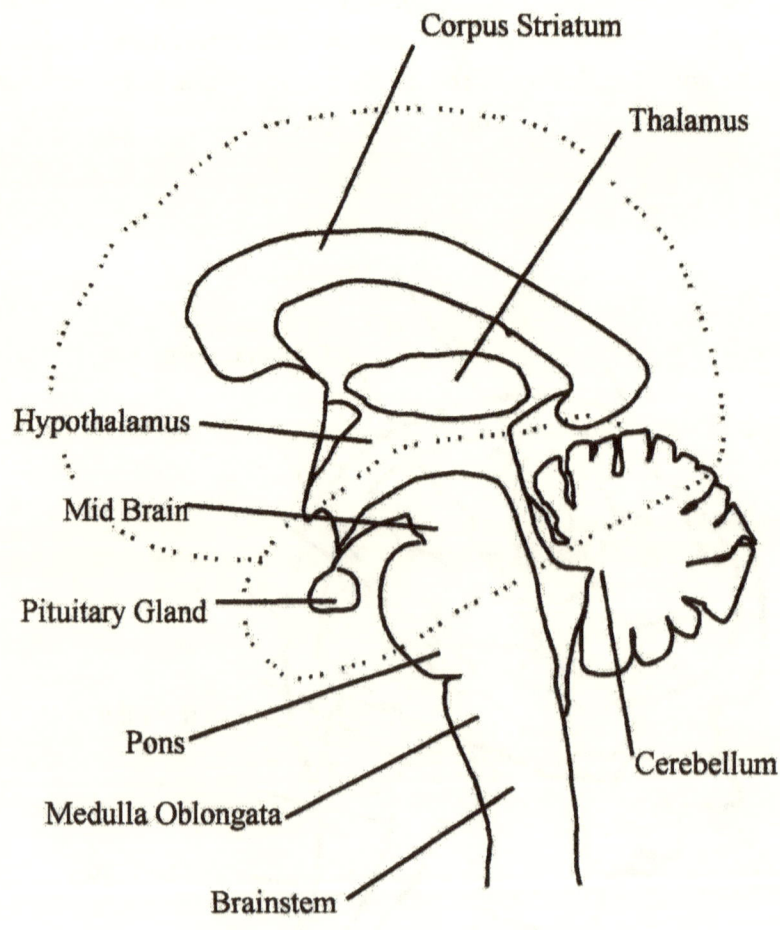

Corpus Striatum

Thalamus

Hypothalamus

Mid Brain

Pituitary Gland

Pons

Medulla Oblongata

Brainstem

Cerebellum

Figure 9.   Brainstem, Cerebellum, Thalamus,
Hypothalamus, Corpus Striatum (left side view).

The corpus callosum is separated into four regions. The most forward region is called the 'Genu'. The Genu curves down and back and
stops, and this is termed the 'Rostrum'. The Genu also curves up and

back and this is termed the 'trunk'. Dipping down, the region of the corpus callosum that is the furthest back is termed the 'Splenium'.

Symmetrical, subcortical masses of gray matter, the basal ganglia, is embedded in the lower parts of each hemisphere. The basal ganglia almost looks like a snake curled up. The head of this snake would be the lentiform nucleus. The neck of the snake would be called the caudate nucleus. The neck curls up and toward the back, then curls around down. The tail of the caudate nucleus wraps around, and then forward to end in a ball called the amygdaloid body.

Two tube-like structures are curved in on themselves. The tube is called the fornix. The forward section of the tubes are balls called the mamillary bodies. The posterior horns of each fornix becomes the hippocampus.

Positioned inside the curved tubes of the fornix is located the thalamus. The posterior and top portion of the thalamus is the pulvinar. There is a right and left pulvinar. In front of the thalamus, on either side, are termed the anterior tubercles. The thalamus itself is a large ball made up of eleven individual nuclei or nerve centers (see Figure 10). The nuclei of the thalamus have been given names on the basis of their position in the thalamus. The thalamic nuclei are known as the Centromedian CM, Lateral dorsal LD, Lateral posterior LP, Medial group M, Medial dorsal MD, Ventral anterior VA, Ventral intermedial VI, Ventral lateral VL, Ventral posterior VP, Ventral posterolateral VPL, and the Ventral posteromedial VPM.

Deep inside the brain, located under the corpus callosum, are four ventricles that hold cerebral spinal fluid. This spinal fluid, a nearly clear liquid, helps bathe and protect the brain. The cerebral spinal fluid drains down to the spine and bathes and protects the spinal cord.

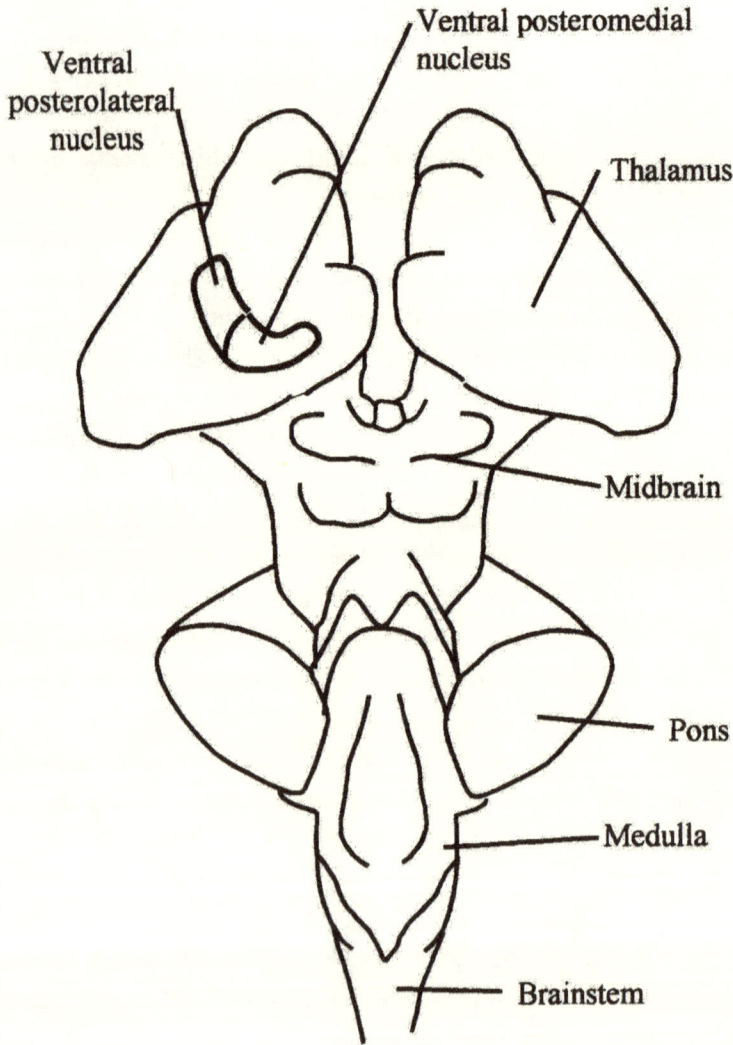

Figure 10. Front view of the core structures of the brain with the thalamus residing on top with two nuclei illustrated on the right side.

The hypothalamus located under the thalamus is made up of a dozen cell clusters called the hypothalamic nuclei. The exact number of nuclei vary depending upon how the subsidiary clusters are recognized. Nerve fibers of the hypothalamus trail into the pituitary gland. The pituitary gland is located at the lower end of the hypothalamus (see Figure 11). The pituitary acts to generate some of the most important chemical regulators in the body called hormones.

Cranial Nerves versus Peripheral Nerves

The nervous system is divided into two functionally diverse systems. The two sets of nerves include the cranial nerves and peripheral nerves.

There are a total of twelve cranial nerves. Cranial nerves originate directly from the brain, generally, from nuclei located in the brainstem. Cranial nerves pass through holes in the skull and travel to a sensor and/or muscle group. The information gathered from these nerves is fed directly back into the brain.

Cranial nerves include eyesight, hearing, taste, and smell. These special senses project nerve fibers from the special sensor to a nerve nucleus located in the midbrain. The nerve nuclei then routs the important data collected by the special senses to the appropriate portion of the cortical brain. The nuclei performs some filtering of the collected data. The most important information is routed to the upper levels of consciousness.

Peripheral nerves are nerves that leave the brain through a large hole in the base of the skull known as the foramen magnum. This bundle of nerves, which descend from the base of the brain, is referred to as the spinal cord. The spinal cord is protected by the bony spine from the base of the skull to the buttocks. Peripheral nerves leave the spine by branches that pass through the holes located between the bones of the spine referred to as a neuro foramina. Once a peripheral nerve leaves the bony spine, they travel to various parts of the body.

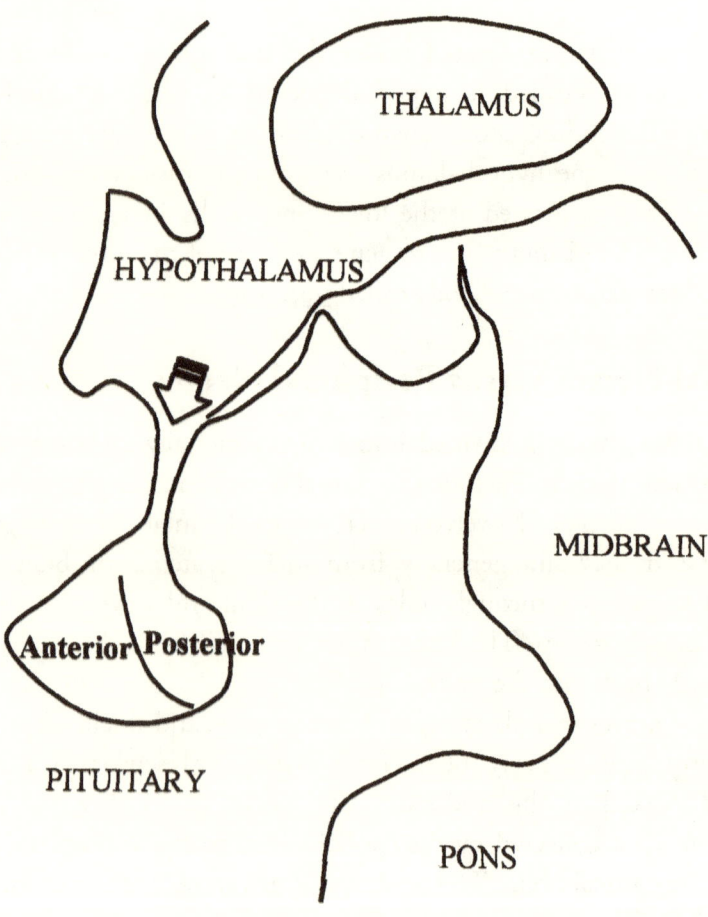

Figure 11. Left side view of the hypothalamus and pituitary gland.

Peripheral nerves act to provide sensor data to the brain. Information that is sent up to the brain includes sensors information regarding temperature, touch or pressure, pain and position (also known as proprioception). Other functions of the peripheral nerves include motor commands to the muscles. The cerebral cortex communicates to the muscles in the body through nerve impulses that pass from the brain through the peripheral nerves to specific muscles. Muscle command signals cause contractions in the muscles that results in movement of the head, limbs or trunk of the body.

The peripheral nervous system (PNS) is further divided into the autonomic and the somatic nervous systems (see Figure 12). The autonomic nervous system, found in all vertebrates, is charged with the control of the body's normal function. The autonomic nervous system frees up the higher cortical centers of the brain from having to take charge and regulate mundane functions of the body such as blood pressure control, heart rhythm, respiration rate, temperature control, etc. The somatic nervous system refers to conscious control of muscles and body organs by the brain, in order to effect interactions with the environment such as finding food, mates, or shelter, and avoiding predators or other dangers.

The autonomic nervous system is further subdivided into the sympathetic and the parasympathetic nervous systems. Since the autonomic nervous system is, much of the time, under subconscious or a form of self-control, to properly regulate the body functions there is a stimulating and an inhibitory input for every organ under autonomic control. This could be thought of as an accelerator and a brake. The sympathetic nervous system causes increases in blood pressure, heartbeat, and breathing. In the case of danger, the sympathetic nervous system reroutes blood from the digestive organs and kidneys to the skin and muscles where the blood is needed to assist in either fighting or fleeing. The parasympathetic nervous system acts to counterbalance the sympathetic nervous system. When the body is safe, digestion, urine production and rest can be effected.

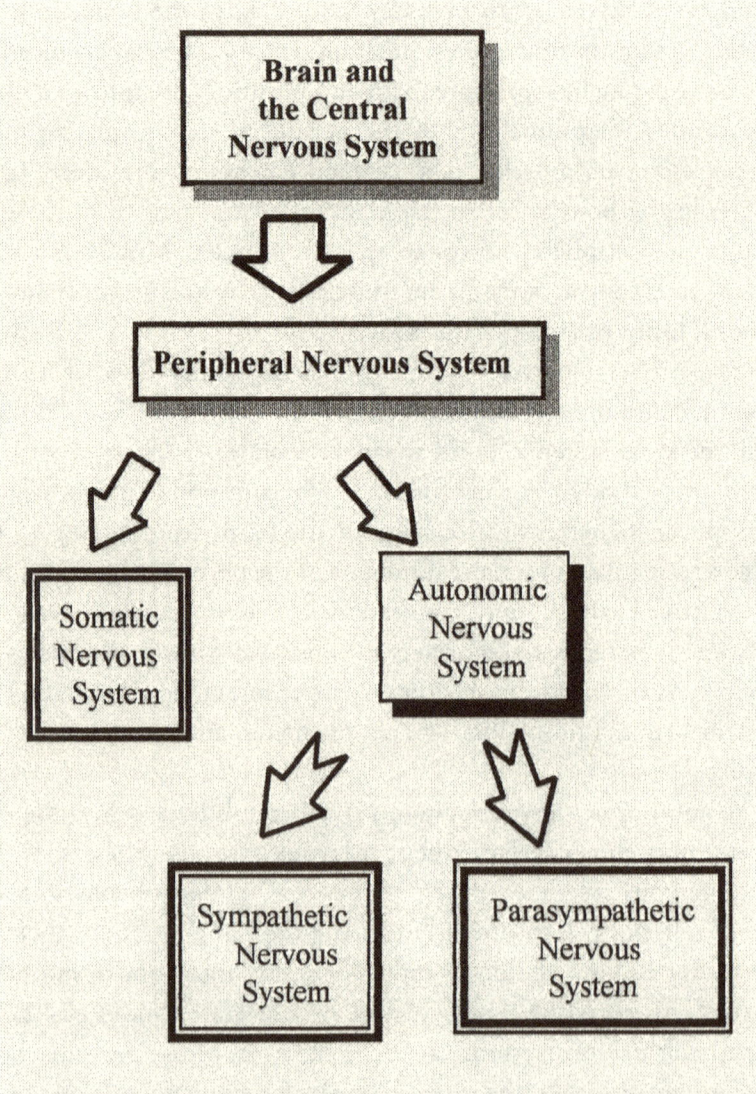

Figure 12. Major divisions of the PNS

11

WHAT IS THE HUMAN COMPUTER?

The human brain is capable of accomplishing a wide variety of tremendous feats. The human brain can communicate both verbally and in written form with other humans using language. Some individuals can master numerous languages and are able to communicate interchangeably with the languages they have learned. The human brain can store vast quantities of information. With practice, the brain can perform various mathematical calculations and accomplish high levels of problem solving. The human brain is able to generate new ideas, cause innovation, master and manipulate the physical world that surrounds it.

Some might say that in calling the human brain a computer would be likened to an insult. The human brain is far more than what we in general, consider a computer to be. In addition, the human brain is capable of so much more than what a personal computer has to offer. The human brain has a sense of self-worth, exhibits emotions, knows how to express love or empathy toward other humans and is capable of reason.

On the other hand, the human brain can be impulsive and irrational, deceiving, plan and effect harm to others and appear at times to operate in a totally illogical and unreasonable manner.

The human brain is comprised of ten billion neurons. It uses chemical and electrical signals to effect information transfer down short and long pipelines of nerve cells. The Intel 8088 microchip, which was fitted into the first IBM PC contained 29,000 transistors. Intel's Pentium III microprocessor contains 9.5 million transistors. The Pentium 4 microprocessor contains 45 million transistors. The prediction is 19 trillion transistors in a single microchip somewhere around the year 2017. The prediction for achievable memory is upwards in the order of 256 gigabytes of memory.

The computer is rapidly evolving in processing power, memory capacity and peripheral capabilities. To believe that future computers won't be able challenge human brain capacity is riveting oneself to old paradigms.

The human brain receives input from central nerves and the peripheral nerves. The human brain analyzes information. Then based on instinct, experience, learned events, emotions, sometimes just pure impulse, the human brain executes movements of muscles and attends to tasks whether they be duties or for pleasure.

The human brain is much more than a computer. But thinking about the human brain in terms of what we know about computers and how they function, can provide valuable insight regarding how our brain operates and what we can do to utilize our brain function in the most optimal manner.

12

NEURONS,
NEUROTRANSMITTERS
AND HORMONES

The desktop computer is restricted to one means of signal transmission, the electron. Electrons pass through the electronic devises that comprise a computer and transfer data by producing or eliminating charges in the computer's devices. The binary code is comprised of a node or circuit being 'on' to represent a mathematical 'one', or the node or circuit not being energized and being 'off' representing a mathematical 'zero'. Present day computers are severely limited by the fact there is only this <u>one</u> mode of data transfer.

Neurons

The brain is comprised of neurons. Neurons are nerve cells that carry messages in the nervous system. Sensory neurons carry information regarding receptor cell data back to the brain. Motor neurons carry messages from the brain in the form of commands to muscles and glands.

Neurons are live cells, and are generally comprised of a cell body, a nucleus, a long body called an axon, and extensions or processes called dendrites. Dendrites receive information and send it to the neuron's cell body. In the brain the dendrites on a single cell may receive as many as 100,000 physical connections from others neurons, though in the human brain, the average is 100 connections. The axon is a projection from the cell body to a target either inside the brain, or to a location outside the brain. There are three types of neurons which include a unipolar, bipolar and multipolar design (see Figure 13).

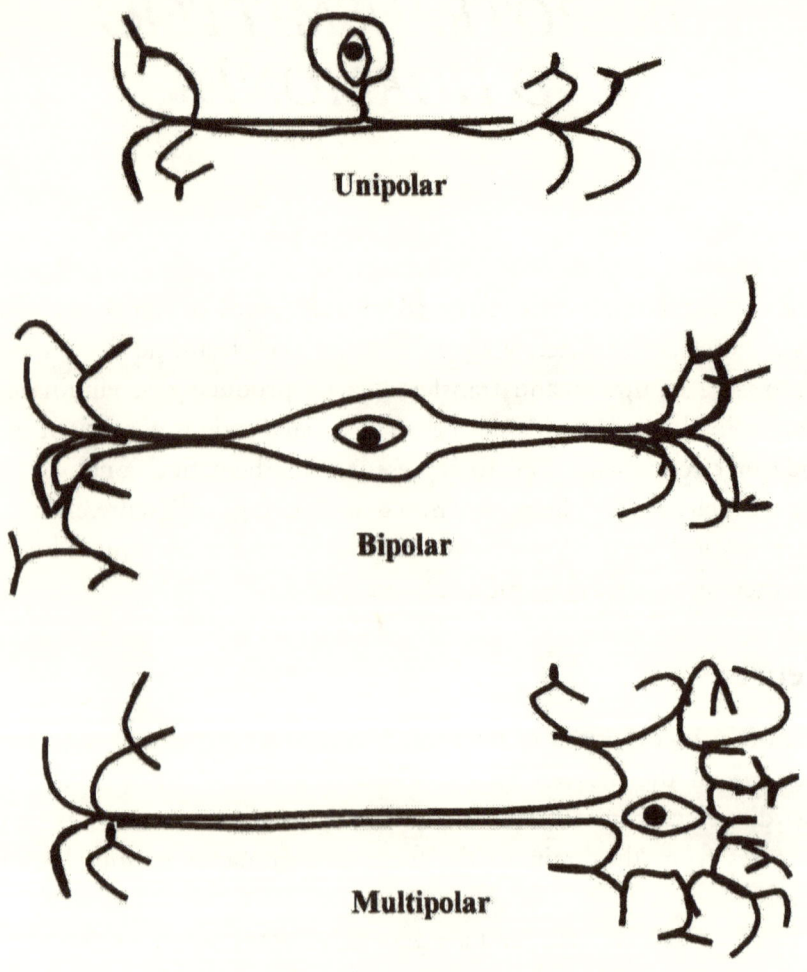

Unipolar

Bipolar

Multipolar

Figure 13. Different types of nerves

Nerve synapse

Where two nerves come in contact to transmit a signal, or the location where a nerve ending terminates on a muscle or a gland is termed a nerve synapse (see Figure 14). Multiple nerve endings may be present at a nerve synapse.

The space between two nerves is termed a synaptic cleft (see Figure 15). In general, a nerve signal is transmitted down an axon from the nucleus (smarts) of the cell. Once the nerve impulse reaches the end of the axon, the nerve ending is stimulated. Vacuoles or sacks containing nerve-transmitting chemicals otherwise known as 'neurotransmitters' are stimulated by the signal. The vacuoles release the neurotransmitter into the synaptic cleft. The neurotransmitter traverses the space between the two neurons and stimulates the nerve ending on the opposite side of the synaptic cleft. The newly stimulated nerve either transmits a chemical signal or an electrical signal to the nerve cell's nucleus.

Neurotransmitters

The brain has a possible 60 different chemical neurotransmitters, which offers the brain a variety of means of transferring data and effecting commands. Neurotransmitters can act to stimulate the transmission of nerve impulses, or they can act to retard or neutralize the transmission of a signal. The effect that a neurotransmitter has on a nerve cell is also in part due to the receptor that the target cell has on the receiving side of the nerve synapse. Receptors may vary in their sensitivity to the effects of the neurotransmitters. Different receptors may cause opposite effects inside the target nerve cell in response to the same neurotransmitter.

Some of the more commonly known neurotransmitters include: Acetylcholine, Dopamine, GABA, Glutamate, Norepinephrine, and Serotonin.

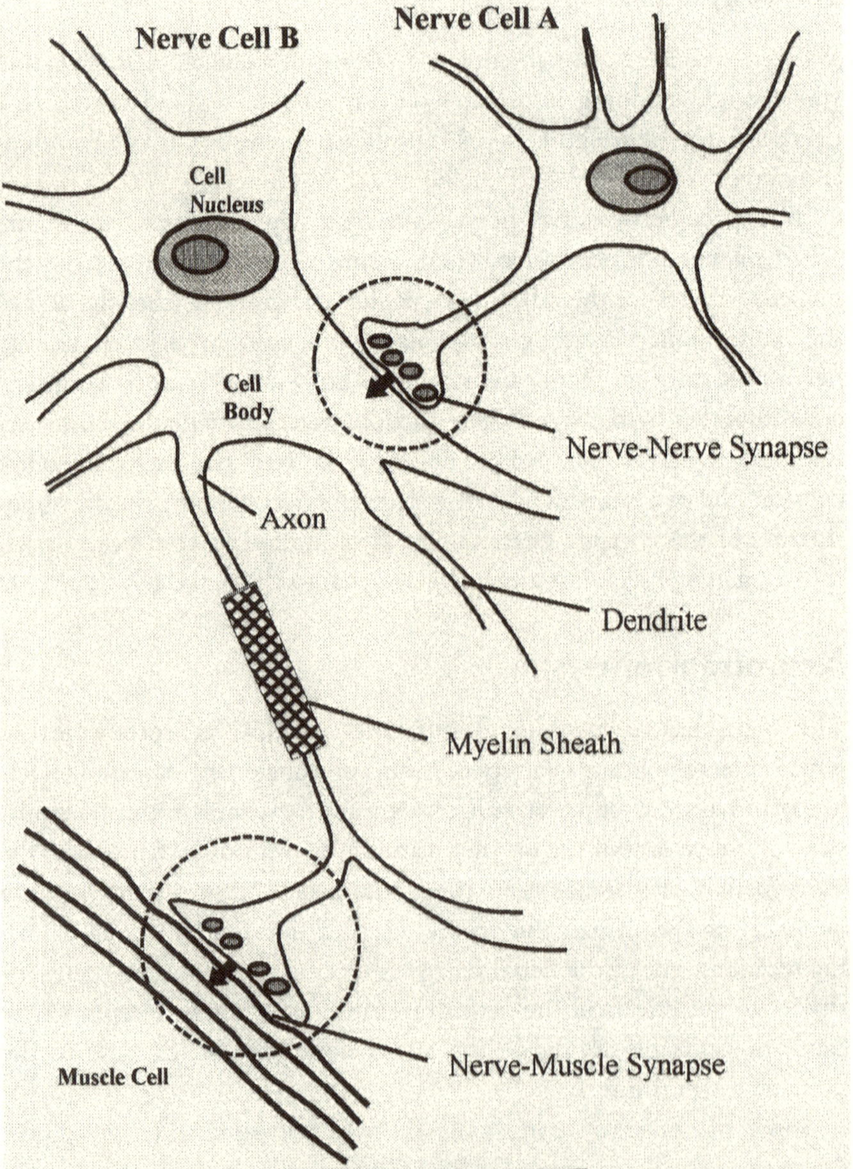

Nerve Cell B

Nerve Cell A

Cell
Nucleus

Cell
Body

Nerve-Nerve Synapse

Axon

Dendrite

Myelin Sheath

Muscle Cell

Nerve-Muscle Synapse

Figure 14.   Nerve Cells

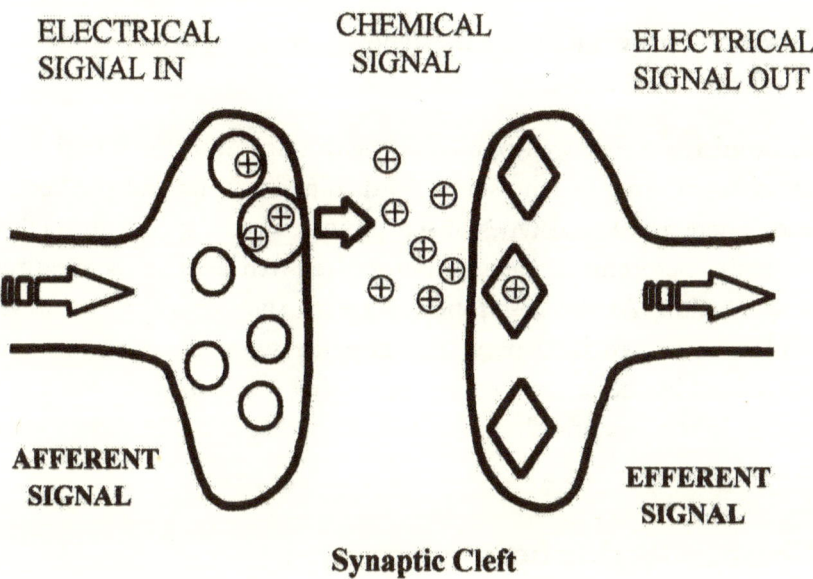

ELECTRICAL
SIGNAL IN

CHEMICAL
SIGNAL

ELECTRICAL
SIGNAL OUT

AFFERENT
SIGNAL

EFFERENT
SIGNAL

Synaptic Cleft

Figure 15. A Nerve Synapsis

Dopamine is one of the most well-known neurotransmitters. Dopamine is synthesized in four major central nervous system pathways. The pathways include nigrostriatal, mesolimbic, mesocortical and the hypothalamic circuits. Low dopamine levels have been associated with the movement disorder Parkinsonism.

GABA or Gamma-Aminobutyric Acid, serves as both a neurotransmitter and a metabolite in normal cell function. GABA is synthesized from glutamate.

Glutamate is a neurotransmitter and a metabolite in cell function.

Norepinephrine is found in the hypothalamus, the lateral tegmentum, and the locus ceruleus.

Serotonin, another well-publicized neurotransmitter, is found in the dorsal raphe nucleus, the spinal cord, hippocampus and the cerebellum. There are several types of serotonin receptors spread throughout the brain. Serotonin uptake inhibitors are often used in the treatment of depression. As the description suggests, such drugs try to prevent serotonin released by a nerve cell into the synaptic cleft from being reabsorbed by the original nerve cell, so that more of the serotonin will remain in the synaptic cleft to stimulate the target cell of certain nerves in the brain.

Neuroendocrine hormones

Where neurotransmitters effect signal transfer between nerve cells and between nerves and muscle cells and between nerves and gland cells, *neuroendocrine hormones* travel in the blood stream from the brain to target organs in the body. Neuroendocrine hormones deliver command signals to organs outside the confines of the brain to indirectly enhance body function.

The Pituitary Gland

The pituitary gland produces a number of neurosecretions known as neuroendocrine hormones. The pituitary is located deep in the center of the brain, just below the hypothalamus (see Figure 16). The pituitary gland receives signals from the hypothalamus, and in response to the stimulus from the hypothalamus, generates various neurosecretions. These neurosecretions are released into the blood stream. These neuroendocrine transmitters leave the confines of the brain, circulating in the blood stream, and stimulate target organs throughout the body.

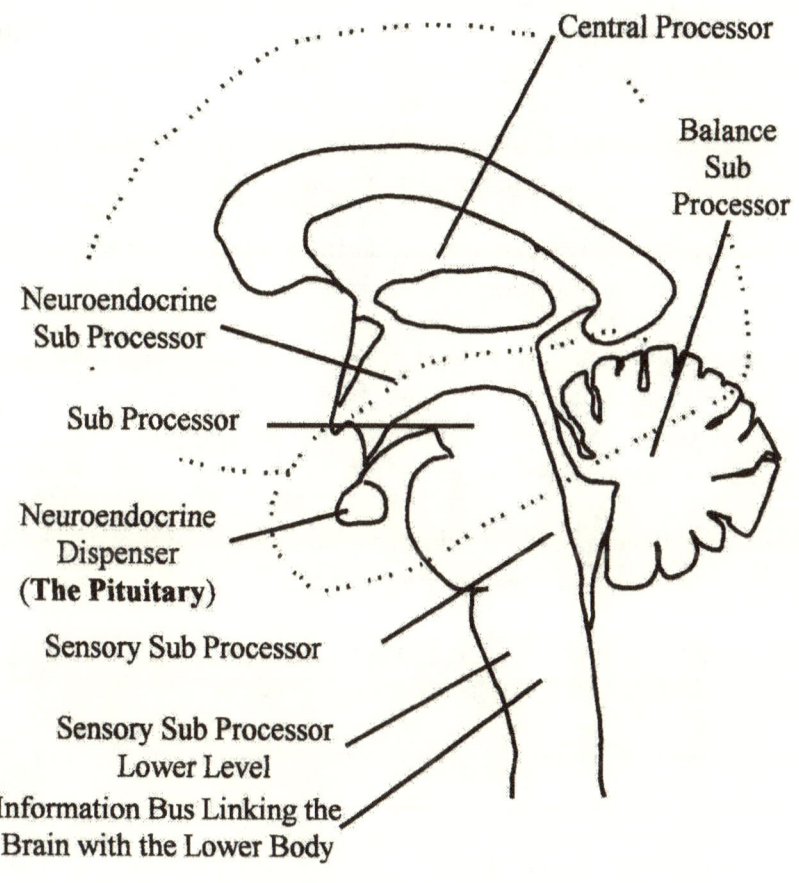

Central Processor

Balance
Sub
Processor

Neuroendocrine
Sub Processor

Sub Processor

Neuroendocrine
Dispenser
(The Pituitary)

Sensory Sub Processor

Sensory Sub Processor
Lower Level

Information Bus Linking the
Brain with the Lower Body

Figure 16. Location of the pituitary gland and
sub processors in the brain (side view).

The pituitary is divided into to two parts. The anterior or front portion of the pituitary produces thyroid stimulating hormone (TSH),
adrenocorticotropin hormone (ACTH), follicle stimulating hormone
(FSH), lutenizing hormone (LH), luteotropic hormone (LTH) otherwise known as prolactin, somatotropic hormone (STH) otherwise

known as growth hormone, and melanocyte-stimulating hormone (MSH). The posterior or rear lobe of the pituitary gland is responsible for producing oxytocin and vasopressin.

Thyroid stimulating hormone (TSH) stimulates the thyroid to produce thyroid hormone recognized as a master hormone in the body.

Adrenocorticotropic hormone (ACTH) stimulates the adrenal glands to further produce hormones.

Follicle stimulating hormone (FSH) stimulates a man's sex organs, the testicles, to produce the male hormone testosterone. Follicle stimulating hormone (FSH) stimulates a woman's sex organs, the ovaries, to produce the hormone estrogen.

Lutenizing hormone (LH) stimulates the ovaries and a developing follicle to produce the hormone progesterone.

Luteotropic hormone (LTH) stimulates the ovary to produce progesterone and stimulates a woman's breast to produce breast milk.

Somatotropic hormone (STH) stimulates growth of the muscles, bone and fat tissue.

Melanocyte stimulating hormone (MSH) stimulate skin cells known as melanocytes.

Oxytocin facilitates contraction of uterine smooth muscle and stimulates the contraction of myoepithelial cells in lactating breasts to generate milk. May act as an important neurotransmitter to stimulate a woman's sexual interest in a man.

Vasopressin is also known as antidiuretic hormone (ADH). Vasopressin regulates renal water retention and vascular smooth muscle tone causing an increase in blood pressure. May act as an important neurotransmitter to stimulate a man's sexual interest in a woman.

Feedback mechanisms in the brain

Pituitary hormone generation is regulated by feedback mechanisms. The hypothalamus generates thyroid releasing hormone (TRH). TRH stimulates the anterior lobe of the pituitary to produce thyroid stimulating hormone (TSH). TSH travels through the blood stream and stimulates the thyroid gland, located in the neck, to produce two thyroid hormones T3 and T4. The levels of T3 and T4 in the blood produce a negative feedback response to the hypothalamus. Too much thyroid hormone and the hypothalamus decreases its production of TRH.

A similar feedback mechanism exists regarding the production of vasopressin (or ADH) in the posterior pituitary lobe. When a person loses water due to sweating, dehydration, vomiting or diarrhea, a smaller volume of blood passes through the blood vessels feeding the supraoptic nucleus located in the hypothalamus. The supraoptic nucleus has nerve endings that travel down and end in the posterior lobe of the pituitary. The posterior pituitary produces vasopressin. Vasopressin is released into the blood stream, travels to the kidneys, and stimulates the kidneys to retain water rather than generate urine. Once the blood volume in the hypothalamus circulation increases to a level considered normal, the production vasopressin is decreased.

Importance of the Pituitary Gland

The pituitary provides the brain an alternate means by which to communicate with the body. Most of the brain's communication with the internal organs and muscle tissues is by electrical impulses that travel up and down nerve routes that exist as information highways comprised of the brainstem and spinal cord. If the brain wishes to move an arm, an elaborate set of command instructions are sent down from the brain via the brainstem, then down the spinal cord to the neck. From the neck, nerve branches exit the cervical spine, travel down the arm to the muscles of the arm. The muscles of the arm move in the manner directed by the nerve impulses sent by the brain. If the nerve sensors of the arm chose to relay information regarding pain, touch, vibration or temperature to the brain, the data is transferred by means of electrical impulses up the arm to the spinal cord, up the spinal cord and brain-stem to the cortical brain where the information is deciphered.

Where most of the command and data transfer is accomplished by means of electrical impulses transferred along nerve fibers, the pituitary gland offers the means for the brain to communicating with the body by chemical signals interjected into the blood stream. Since blood is actively circulated throughout the body by the pumping action of the heart, and since blood reaches nearly all of the tissues of the body, the pituitary offers a means of the brain to interact with a single target organ or numerous organ systems, in a broad sense, all at one time.

The pituitary gland would be analogous to a computer's CPU having the capacity to communicate directly with the other internal devices of the computer by means of light, sound, smell or vibration, in addition, to the electrical impulses sent along the internal data buses.

The pituitary gland and the essence of the manner in which it works, offers a powerful concept that challenges human computer architect designers to incorporate alternative means of data and command transfer into the workings of future generations of desktop computers. The pituitary gland idealizes that human brain has been designed in a manner that is 'outside the box'. This should be studied in more depth, to assist us in creating computers that are constructed of designs that exist outside the current box.

13

THE HUMAN COMPUTER'S CPU, MEMORY & DATA BUSES

If for a moment, we were to consider the human brain as a type of computer, how would it be constructed?

The human computer surely would need the basics of what we would expect from a desktop computer. The human computer would need a Central Processing Unit, Memory devices, an Arithmetic unit and data buses to connect the internal parts of the brain with one another.

The CPU of the brain would be in the center of the brain making it most accessible to all functioning portions of the brain. The Human Central Processing Unit (HumCPU) is comprised of the Lentiform Nucleus and the Diencephalon. The Lentiform Nucleus is made up of the Globus Pallidus I & II and the Putamen. The Diencephalon is comprised of the Thalamus, the Hypothalamus, the Subthalamus and the Epithalamus. Other structures that make up the HumCPU include the Amygdala, the Caudate Nucleus, the Lateral Geniculate Nucleus and the Medial Geniculate Nucleus. (See Figures 17, 18, 19, 20 and Figure 21)

Memory would be located in several parts of the brain. One would need memory space to store visual memory files, auditory (sound) files, fact files, olfactory (smell) memory files, tactile memories, somatic motor files. Somatic motor memory files are muscle functions that are repeated repetitiously such as required to participate in a sport, play an instrument, or perform a work related task.

Data buses would transfer information from elements of the body to the brain. Data buses would transfer information from the front to the back of the brain, and from the right hemisphere to the left hemisphere in the brain.

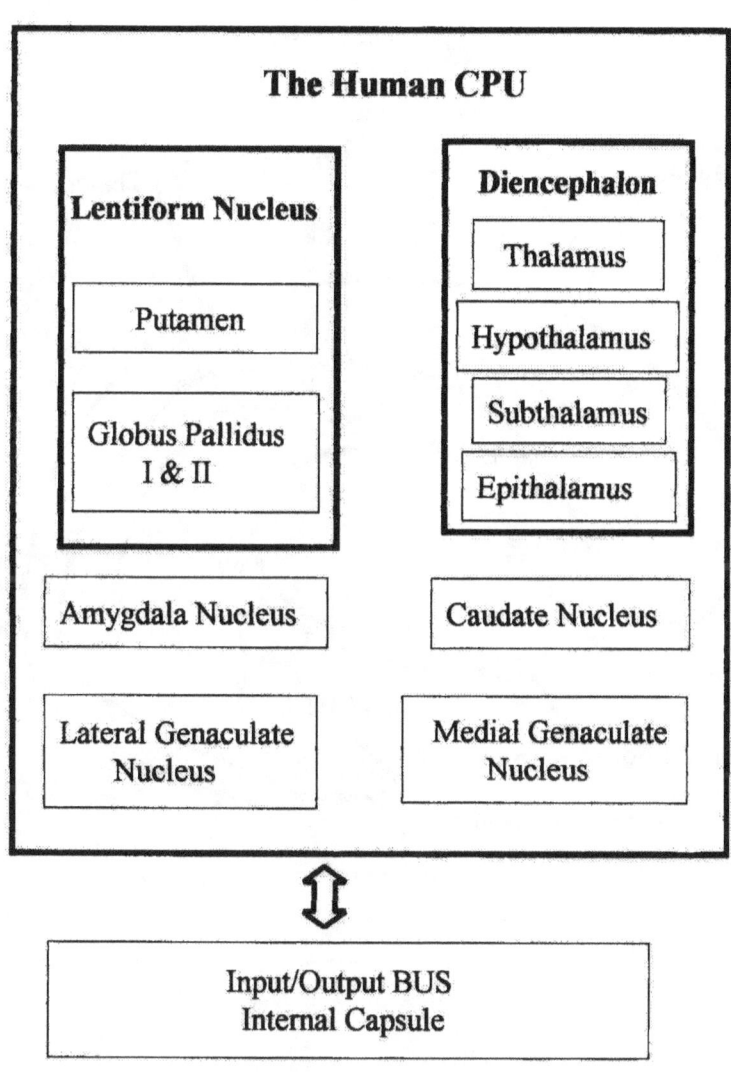

Figure 17. Components of the Human CPU

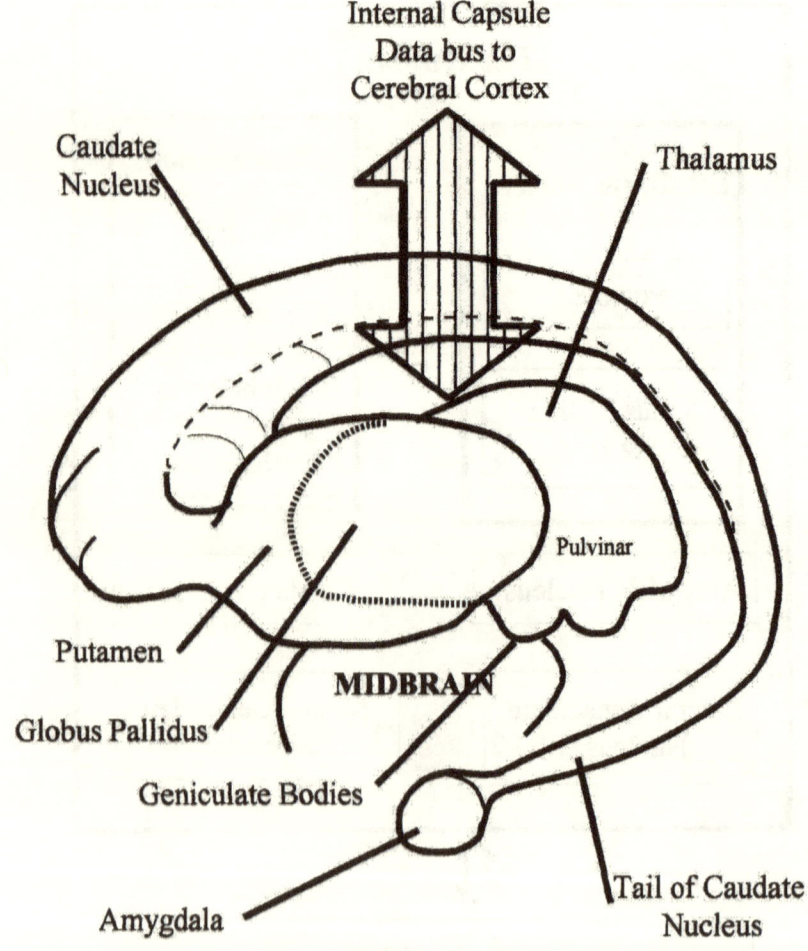

Figure 18. Human Central Processing Unit including the diencephalon and the lentiform nucleus (left side view).

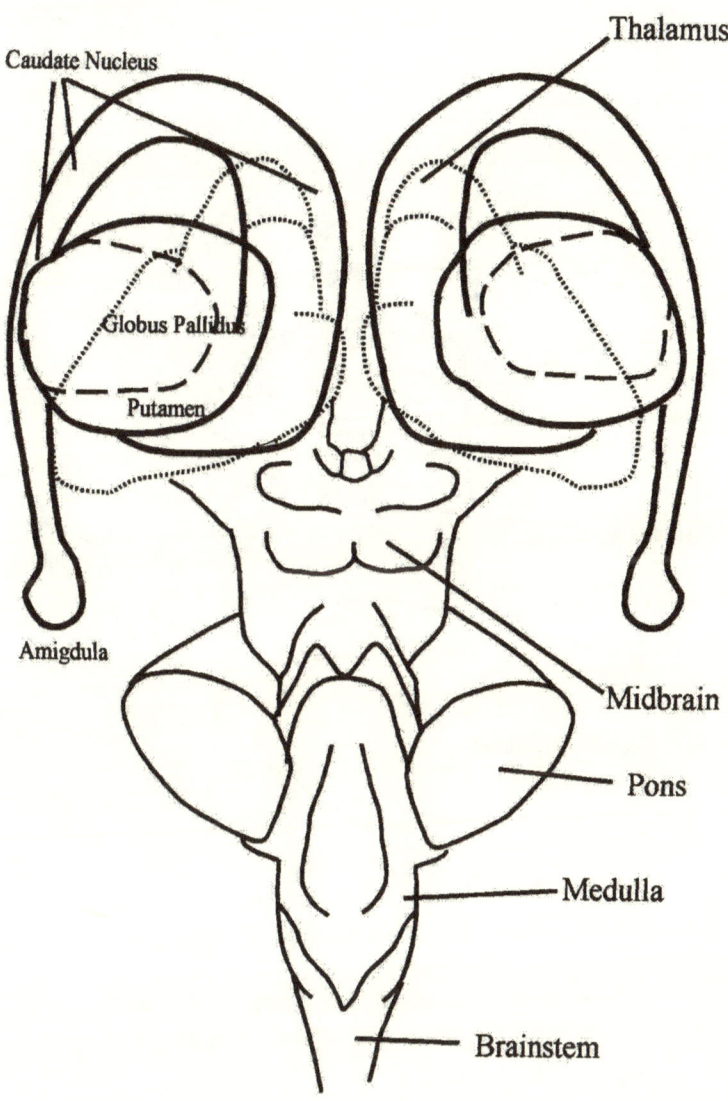

Figure 19.   Central Processing Unit, Midbrain, Pons, Medulla Oblongata, and Brainstem (front view).

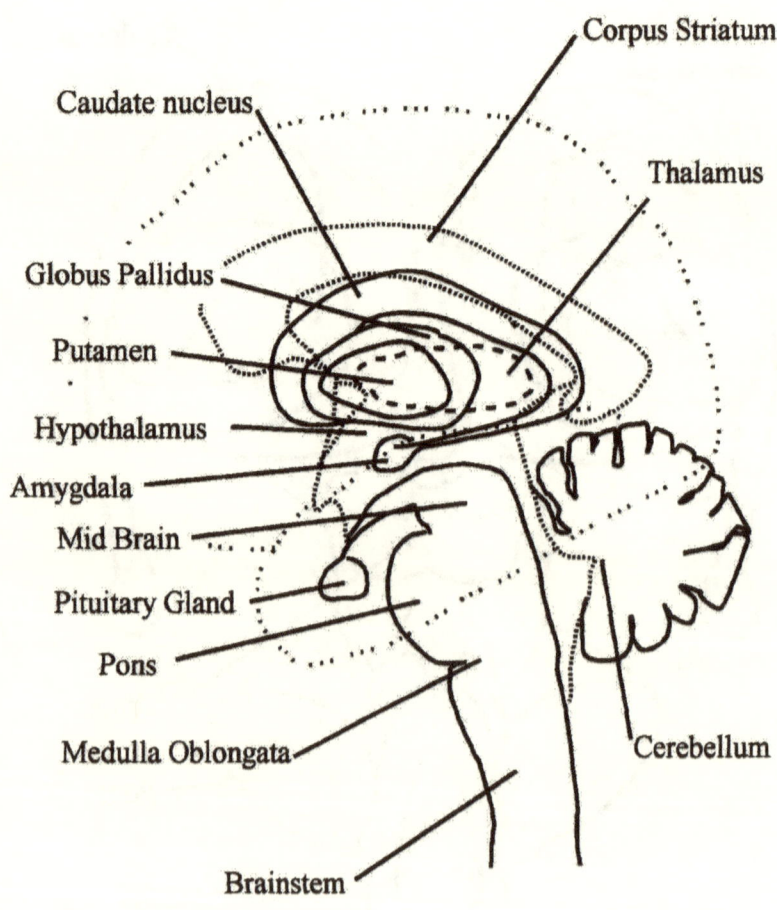

Figure 20. Corpus Striatum, Central Processing Unit, Midbrain, Pons, Medulla Oblongata, and Brainstem (left side view).

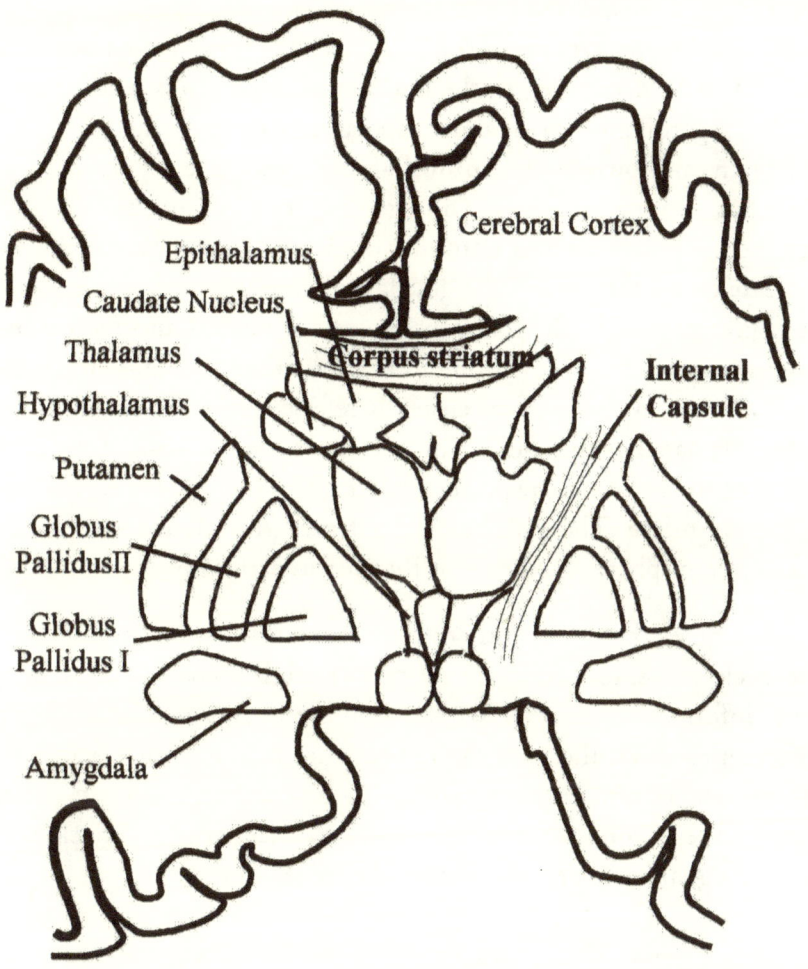

Figure 21. Front view of a slice through the brain showing the structures surrounding the Thalamus (viewed as if you cut the brain down across the midline, referred to as a sagittal cut).

Several data buses have been identified. The bus from a parietal lobe of the brain down to the muscles in the limbs is called the **pyramidal** tract (note there are two pyramidal tracts, one for the right side of the body, one for the left side of the body and these two nerve tracts cross, referred to as 'decussate', in the brainstem). A data bus sending information regarding fast pain and temperature data from the limbs up to the brain is known as the **spinothalamic tract** (see Figure 22). The data bus sending information regarding tactile (touch), vibration, proprioception (limb position and motion) include the **gracile and cuneate tracts** that end in the gracile and cuneate nuclei, then transfer their information to the thalamus by way of **medial and lateral lemniscus tracts**. The information bus from the thalamus up to the higher cortical areas in the cerebral cortex is referred to as the **internal capsule**. The data bus that carries information from the front of the cortex (front of the brain) to the posterior cortex (back of the brain) is known as the **cingulate**. The data buses that travel from right hemisphere to the left hemisphere are known as the **corpus striatum**.

Data buses that go directly to the brain without having to come up the neck and spinal cord are termed **cranial nerves**. Cranial nerves send information directly to the brainstem and receive command information from the brainstem. There are twelve cranial nerves that act to send information directly to the brain. The brain uses cranial nerves to send commands to muscles in the face, head and neck, without the signal having to go through the spinal cord.

Cranial nerves send input data from sensors along first order neurons. These first order neurons end in nuclei located in the brainstem (see Figure 23). From these cranial nerve nuclei the information is transmitted up into the cerebral cortex of the brain by second and possibly third order neurons. See Table 1 for a list of the twelve cranial nerves.

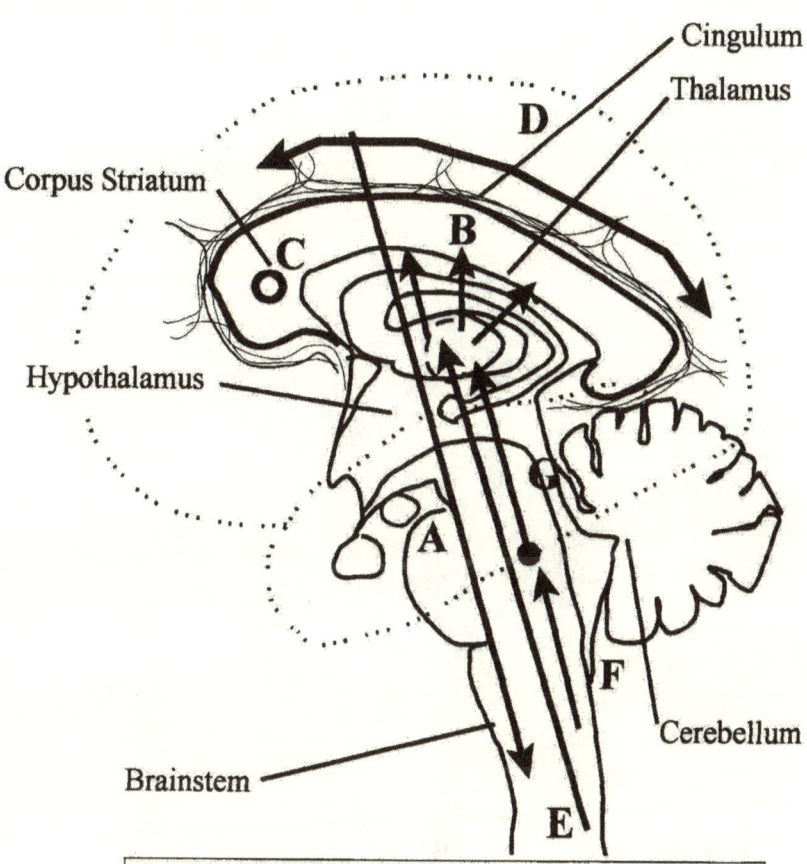

Figure 22. Major information buses of the brain
A: Pyramidal Tract (down)
B: Internal Capsule (up/down)
C: Corpus Striatum (side to side)
D: Cingulum (front to back)
E: Spinothalamic Tract (up)
F: Gracile & Cuneate Tracts (up to G)
G: Medial & Lateral Lemniscus Tracts (up)

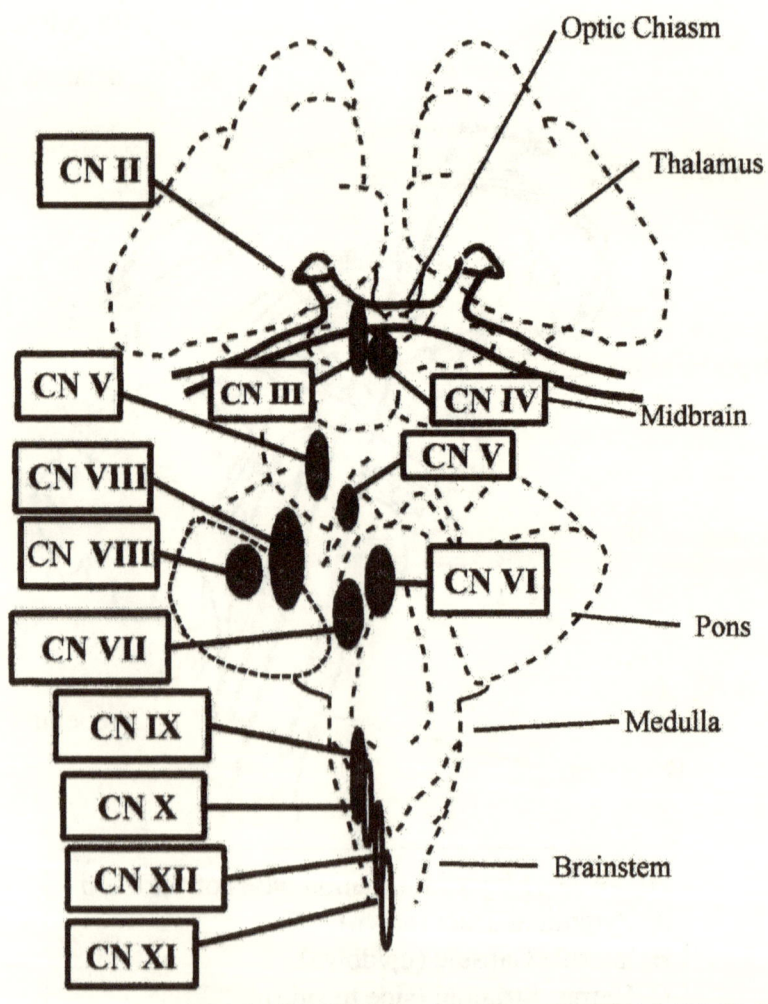

Figure 23. Front view of the lower brain structures and the location of the cranial nerve (CN) nuclei.

Table1. A list of the twelve cranial nerves and a brief description of their functions.

Number	Cranial Nerve	Function
I (1)	Olfactory	Smell
II (2)	Optic	Sight
III (3)	Oculomotor	Movement of the eyes
IV (4)	Trochlear	Movement of the eyes
V (5)	Trigeminal	Face innervation
VI (6)	Abducent	Movement of the eyes
VII (7)	Facial	Face innervation
VIII (8)	Vestibulocochlear	Balance and hearing
IX (9)	Glossopharyngeal	Throat/tongue movement
X (10)	Vagus	Throat/tongue movement
XI (11)	Accessory	Throat/tongue movement
XII (12)	Hypoglossal	Tongue movement

14

HUMAN PERIPHERAL DEVICES AND FEEDBACK LOOPS

The human computer has a number of peripheral devices to work with. These peripheral devices include those that exist inside the brain and those that exist outside the brain.

The Human Body's Organ Systems

Heart: Four chambered pump located in the chest, that acts to push blood throughout the arterial system of the body.

Lungs: Inflatable air bags located in the chest, that facilitate oxygen being absorbed from the air by the red cells in the blood and the expulsion of carbon dioxide into the air.

Liver: Large solid organ in the right side of the abdomen that detoxifies drugs we ingest, harmful agents the body produces, filters the blood and produces beneficial proteins.

Spleen: A solid organ on the left side of the body that filters the blood to remove bacteria or other infectious material.

Kidneys: One on either side, located in the mid back region, guarded by the lower ribs, filter the blood and excretes toxins and excess water in the form of urine.

Adrenals: Small, sliver shaped glands that sit above the kidneys and produce vital hormones, including cortisone.

Stomach: A bag, located at the lower end of the esophagus in the upper abdomen, that produces acid and initiates digestion of food.

Intestines: A long tube, comprised of the small intestine and the colon, extending from the lower stomach to the anus, that absorbs nutrients from the food that has been eaten and digested, and secretes waste materials.

Bladder: Stores urine produced by the two kidneys, until the urine is of sufficient quantity to be expelled.

Ovaries: (Women only) Two small solid organs about the size of a small egg, located in a woman's pelvis, that produce estrogen and progesterone in varying amounts during the course of a month in the non-pregnant woman, the levels of the hormones being responsible for fertility and the menstrual period a woman experiences on a monthly basis.

Uterus: (Women only) An organ designed to accept a fertilized egg cell and create a nourishing and protected environment for the egg cell to grow into a fetus.

Testicles: (Men only) Two solid organs the size of a small egg, located in the scrotal sack, that produce the male hormone testosterone and produce sperm cells.

Prostate: (Men only) A gland that sits between the bladder and the base of the penis, produces a portion of the fluid used in the ejaculate the male produces during sexual intercourse.

Muscles: Long strands of cells anchored on two different bones that contract when stimulated by an electrical signal delivered to the muscle by a nerve.

Immune System: Comprised of T-Cells, which act as roving police, and B-Cells that exist in lymph nodes and the spleen, that remember past infections, and produce antibodies against infections when stimulated.

Lymph nodes: Small solid structures scattered throughout the body, that act as nodal points (similar to satellite police stations) where infectious agents such as bacteria and viruses are identified and the immune response is mounted against the infectious agent.

The human brain includes the subconscious and the memory unit. The subconscious interacts with the frontal lobes of the brain to work in a partnership. Our consciousness works on higher level thinking, which makes each of us an individual, while the subconscious toils with all of the lower level thought processes that are required to make the body run.

The subconscious is a liaison between the consciousness, the memory unit, the cranial nerves and the peripheral organs and muscles. The subconscious also interacts with the emotional unit in the brain, the sexual drive unit in the brain, and the logic unit in the brain.

The peripheral devices outside the brain include internal and external sensory devices, sound generation by the vocal cords, muscle function, and other bodily function such as devices to regulate body temperature, body blood sugar, blood pressure, body water balance, energy conversion and so forth. The external sensory devices include optics, audio perception, sense of smell, sense of touch and pain, and the sense of taste.

Humans have a complex series of sensors that allow us to investigate and monitor the environment that immediately surrounds us. Our skin is laced with a series of nerve endings including pain receptors, temperature sensors, and proprioception sensors. Proprioception sensors tell the brain where a part of the body is located in space. That is, proprioception sensors tell the brain where your hand is located such as either scratching your head, laying in your lap, at your side, or slapping at a mosquito on your leg. Other sensors include the special senses of sight, smell, taste, and hearing. It is the coordinated efforts of the peripheral and special senses that keep our bodies updated to changes that occur in the environment around us.

If we walk outside in our bare feet, and step on black top on a sunny day, the temperature and pain fibers in our foot will sense the heat of the pavement and send a signal to the brain. If the temperature of the pavement exceeds that which is considered safe by the temperature regulator in the brain, then a signal will be sent down to the foot to remove itself from the pavement and seek a safer, in this case a cooler footing. If the encounter with an object is sudden, and extremely hot or cold, the temperature sensor will send a signal along the nerve root to the spine. At the spine nerve junction in the spinal cord the information will be routed up to the brain so that the conscious brain will be aware of the potential danger. At the spinal cord, if the signal is strong enough from the peripheral sensor, the spinal cord will automatically send an efferent signal back to the limb where the temperature variation was detected, to instruct the limb to remove itself from the source of the temperature variation. This is why, if our hand detects something hot, such as a flame, our hand is oftentimes removing itself from the heat source before we consciously recognize the danger. This provides a measure of safety to the limbs to prevent tissue destruction.

15

PUTTING THE HUMAN COMPUTER ALL TOGETHER.

As was discussed in earlier chapters, the desktop computer is comprised of inputs such as the keyboard, the mouse, a scanner, the phone line, a scanner, a digital camera, and a responsive video-touch view screen or video-touch keypad (see Figure 24). The major computer components include the central processing unit (CPU), the hard drive, and power supply. Input/output information devices include the compact disc (CD) drive unit, the DVD drive unit, the 3 inch floppy drive, cassette tape drive, a hand-held mobile computer unit with a cradle, a dedicated Internet access cable, infrared port and the modem. Output devices include the printer, the video screen, the speakers (see Figure 25). The central processor and interrupt control are illustrated in Figure 26. An overall schematic diagram of the desktop computer is presented in Figure 27.

So how does the human brain compare to the construction of the desktop computer: *Remarkably similar!*

To begin to understand the human brain and how it functions, you have to do what you did with the computer: You must rationalize that the human brain, like the desktop computer, is **not** a black box with magical powers. The human brain exists in a physical universe and something has to make it work.

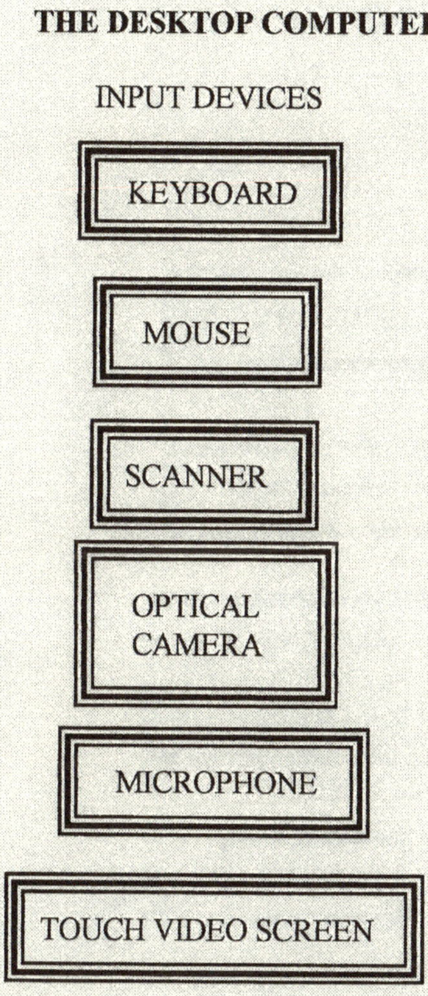

THE DESKTOP COMPUTER

INPUT DEVICES

KEYBOARD

MOUSE

SCANNER

OPTICAL
CAMERA

MICROPHONE

TOUCH VIDEO SCREEN

Figure 24. Basic Inputs to the Desktop Computer

THE DESKTOP COMPUTER

OUTPUT DEVICES

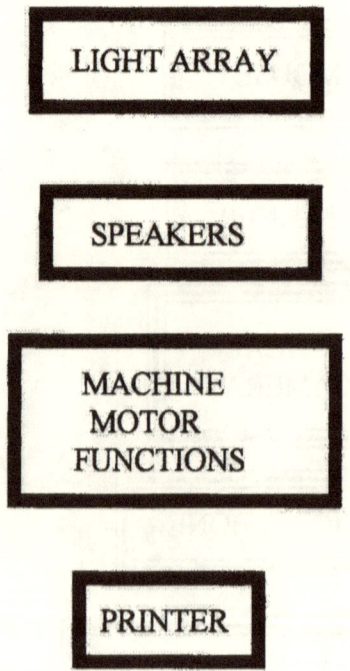

Figure 25. Basic Outputs of the Desktop Computer

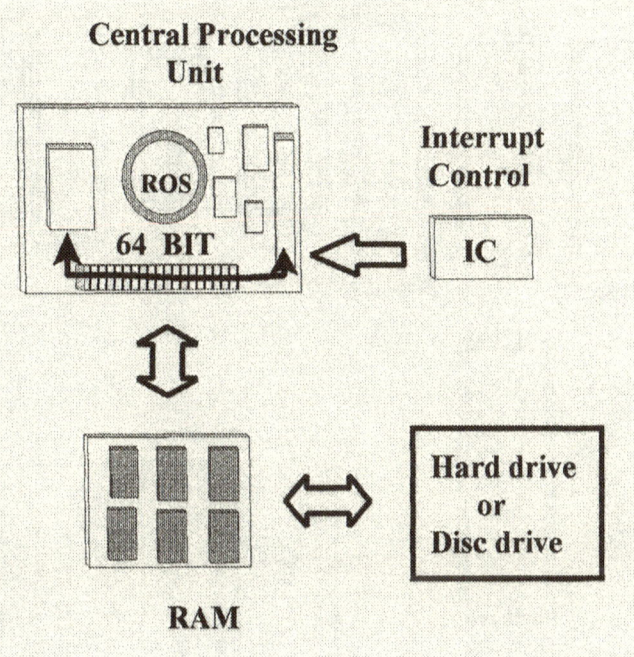

Figure 26. Central Processing Unit and
Interrupt Control

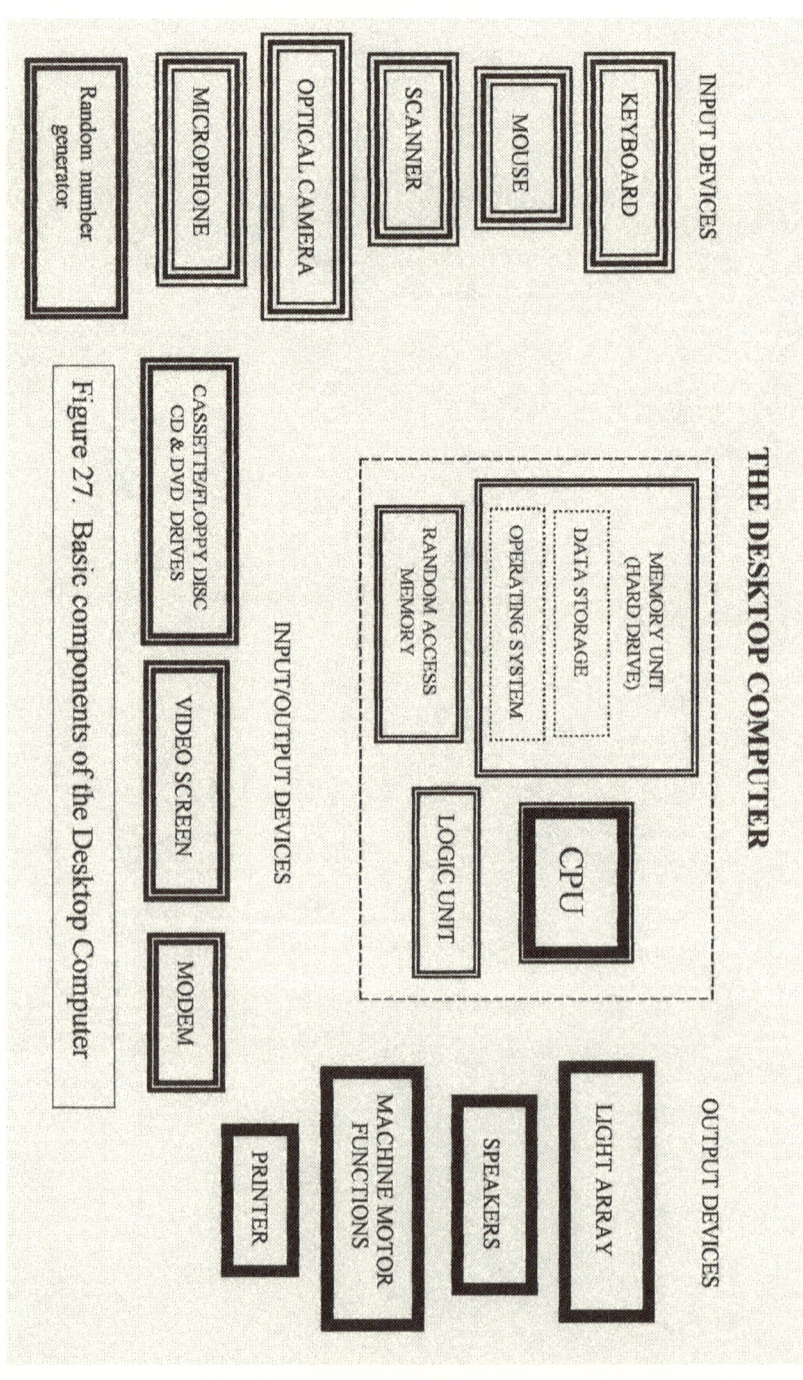

Figure 27. Basic components of the Desktop Computer

Putting the concepts of the human computer, all together, is slightly more challenging than putting it all together for the tower computer. You cannot just pick up a screwdriver and have at it. It is much more difficult to study what is inside the human brain, than what exists inside the personal computer. The human nervous system can be mapped out and when you consider the possibility, appears to be very similar to a desktop computer system.

The brain sits on top of the spinal cord. The spinal cord is comprised of tracts of long nerves that send signals down to the body and receives data input from sensors located all over the body. Fast pain and temperature sensors send their data up to the brain by way of the spinothalamic nerve tract. The vibration, tactile (touch), proprioception (position sense) generally located on the surface of the skin send their data up to the brain by way of the Gracile and Cuneate nerve tracts. The rudimentary subconscious brain controls the body by interpreting the sensor inputs and adjusting position, moving muscles or changing levels of activity or types of activity in order to keep the body comfortable. The higher-level cortical brain, is aware of the elements in the surrounding environment, and has wants and desires.

The higher-level conscious brain instructs the more rudimentary, subconscious brain, to effect body function in order to satisfy the wants and desires it has. This is similar to a computer where the user of the computer interfaces with the computer through a peripheral device such as a keyboard or a microphone connected to voice activated software. Where a computer user may type in or say 'print screen', the software the user is interfacing with sends instructions to the CPU, and the CPU interprets the instructions on at the level of the rudimentary machine code software, then effects the proper sequences of memory and printer functions in order to produce the printed material the user is requesting. Once the function is complete the computer waits for the next task.

So, how is it that the human brain functions as a computer? Typically, we consider that the brain is divided into white and gray matter. Most of us are aware that the white matter suggests intelligence and gray matter is associated with a more rudimentary portion of the brain. Unfortunately, this concept is relatively correct, but this is also where most of our understanding ceases.

The white matter can be associated with higher levels of brain function and consciousness, due to the simple fact that humans regard themselves as the most intelligent life on the planet and our brains contain the largest volume of white matter amongst the various animal life that inhabit the planet.

If we review what is known of the anatomy of the human brain (see Figures 28) and associate this with the internal structure of the present day desktop computer (see Figure 27) some associations can begin to be etched out.

Major Divisions of the Cerebral Cortex

FRONTAL LOBES
PARIETAL LOBES
OCCIPITAL LOBES
TEMPORAL LOBES

FRONTAL BRAIN LOBES

The frontal portion of the brain, referred to as the Frontal Lobes (right and left), are thought to be associated with an individual's ability to reason, expression of emotion, and personality (see Figure 29). The frontal lobes of the brain therefore, must contain what we would consider our consciousness, what we consider our 'person' or 'individuality'. In a computer, this would be considered the area of the hard drive that contains the interface program software that a computer's user would interact with to run the computer. In other words, this would be where the higher-level operating system would be stored.

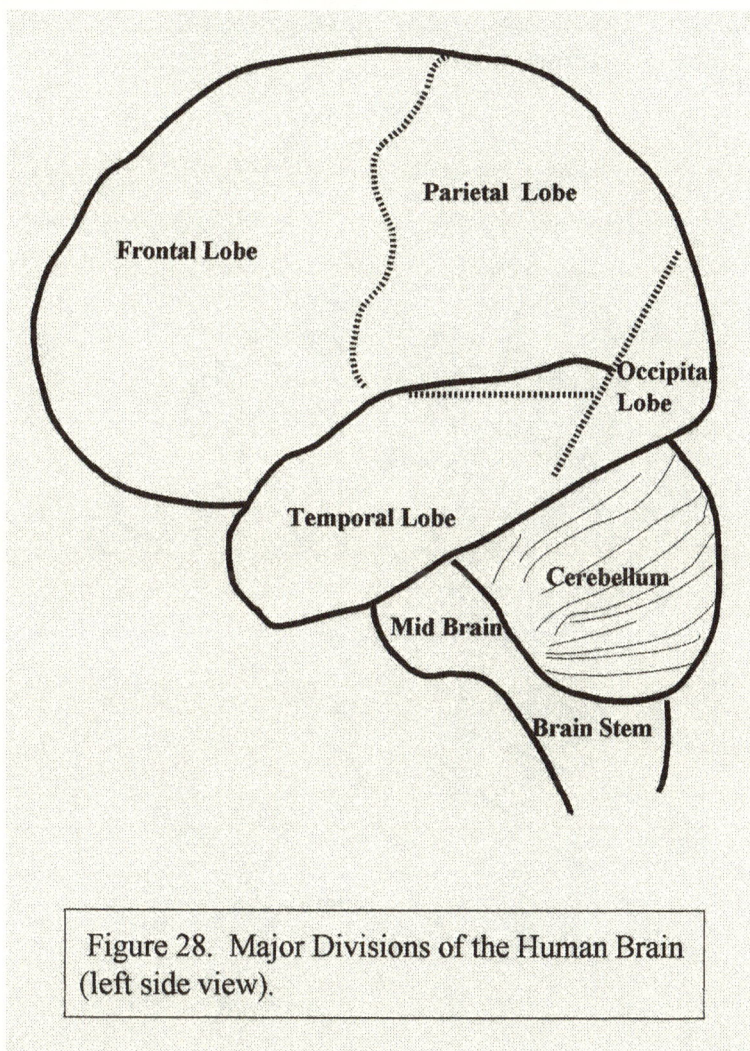

Figure 28. Major Divisions of the Human Brain (left side view).

THE PARIETAL LOBES

The desktop computer generally does not have to worry about movement, except for an accidental push or nudge off the top of a work station, at which point the unit might crash to the ground. The desktop computer does have a mathematics unit that is dedicated to performing mathematical calculations, when necessary, while the CPU spends its time attending to other duties.

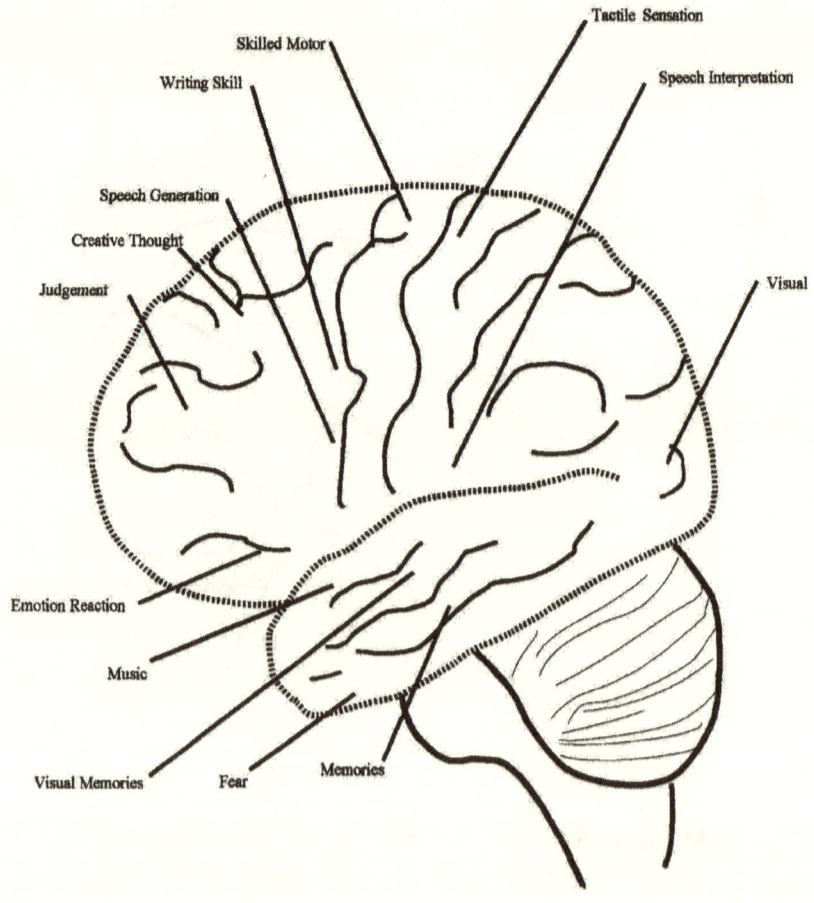

Figure 29. The Human Brain Divisions of Brain Function in the Upper Cortex (left side view).

Humans, like other animals, have a necessity to move within their environment to seek food, shelter, water, and other objects of their desire. Movement is a very complex human function requiring numerous inputs of data and outputs of coordinated commands. To move an arm requires spatial sensory input, termed proprioception, in order for the brain to know where the arm is located in space and what position the arm is currently in; is the arm straight, is the arm bent at the elbow, is the arm in the back, at the side or in front of the body. Once the location of the arm is established, then a plan must be generated in order to change the position of the arm to the new location that the conscious or subconscious brain wishes the arm to be located. When we are awake and the conscious brain is in control of the body's function, movement of the arm and other three limbs is generally under conscious control. When we are asleep, the subconscious brain may move the arm periodically in order to prevent the arm from being crushed by the weight of the body, or to flick at an insect trying to bite us.

Movement of the body, much less the movement of a single limb, requires orchestrating and enacting very complex functions and demands a significant amount of computer processing power. The parietal lobes of the human brain are where sensory input and motor control of the limbs of the body are developed. The parietal lobes act as a separate computer processor dedicated to the proper function of body motion.

Motion of the body and limbs of the body are designed by the parietal lobes and the generated commands are passed down from this higher cortical processing area of the brain into the diencephalon. The diencephalon then routes the movement instructions to the proper spinal cord tracts and sends nerve impulses down the spinal cord to various levels of the spine, which results in nerve impulses being sent down the tracts of peripheral nerves to the muscles that will effect either movement of the body or desired movement of a limb.

The parietal lobes act as a separate processing unit, to prevent the thalamus from being bogged down with designing and implementing movement of the human body, much like a mathematics unit acts as a separate processing unit to free up the CPU from having to be bogged down by performing mathematic calculations or a video card performs translations of digital data to imagery which is then transferred to the video screen so that the CPU isn't tied up doing such labor intensive tasks.

THE THALAMUS

The thalamus is not apart of the cerebral cortex, but is vital to the cortical lobes, by acting as an information transfer center. The thalamus sits at the top of the brainstem, above the midbrain. This area of the brain is termed the diencephalon. The thalamus is where at least eleven nerve nuclei reside. The thalamus acts as a switching station for nerve impulses coming up the brainstem, and nerve impulses passing down the brainstem from the cerebral cortex. The thalamus is most likely a significant part of the central processing unit of the brain. The thalamus receives instructions from the higher levels of the brain and helps execute those processes. The thalamus is also where the inputs from the body's sensors are routed up to the appropriate portions of the brain. The thalamus controls input and output of nerve impulses, much like a computer's central processor, controls the input, output and flow of commands and data through the desktop computer. The thalamus is a component of a larger structure termed the diencephalon.

OCCIPITAL LOBES

The occipital lobes are positioned at the rear of the brain. It is in the occipital lobes that the interpretation of sight occurs. Light enters the lens of each eye. The retina at the back of each eye translate light into nerve impulses. The nerve impulses are transmitted from the eyes at the front of the head to the occipital lobes at the rear of the head by means of a data bus termed the optic nerve. The optic nerve relays the light information to the optic radiation, individual nerve projections, that spread across and connect to different portions of the occipital lobes. The occipital lobes interpret light images for the conscious brain.

TEMPORAL LOBES

The temporal lobes sit to the outside of the parietal lobes and connect to the parietal lobe and occipital lobe. The temporal lobes act as the part of the brain that facilitates long-term memory storage. The temporal lobes also participate in the recognition and storage of music, and the recognition of fear.

The two halves of the brain are considered hemispheres. If a person uses the left side of the brain to create speech, they are considered left-brain dominant. The left side of the brain controls the right arm and right leg, therefore, those individuals that find themselves stronger on the right side are considered to use the left side of their brain more than the right side of their brain. This preconception may be a half-truth. In left side dominant individuals, intelligence, calculating skills, and rationalization skills are facilitated by the left frontal lobe, language comprehension and articulation are formulated in the left lower parietal lobe, and hand writing skills are generated in the left upper parietal lobe (see Figure 30). On the other hand, intuition and geometric skills are facilitated by the right frontal lobe, drawing and art skills are created in the right parietal lobe, speech intonations and gestures as well as facial recognition skills are also facilitated by the right side of the brain. Right brain dominant individuals function just opposite of the above description. There are a number of structures that comprise the human brain. As an introduction to the Human brain, brief descriptions follow as:

RIGHT HEMISPHERE OF THE BRAIN

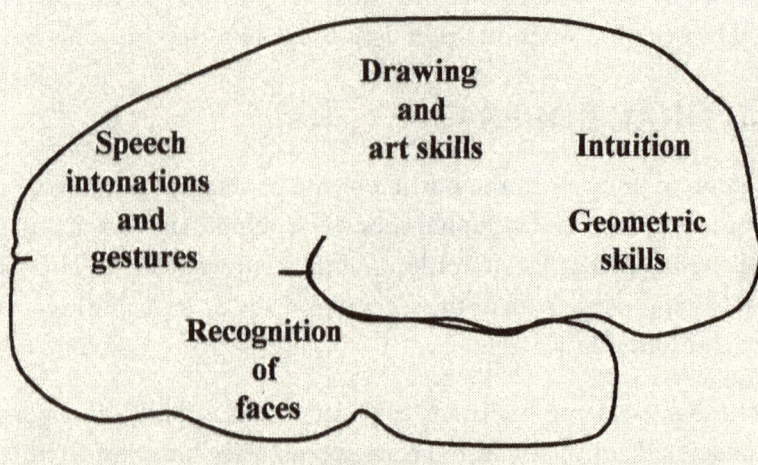

LEFT HEMISPHERE OF THE BRAIN

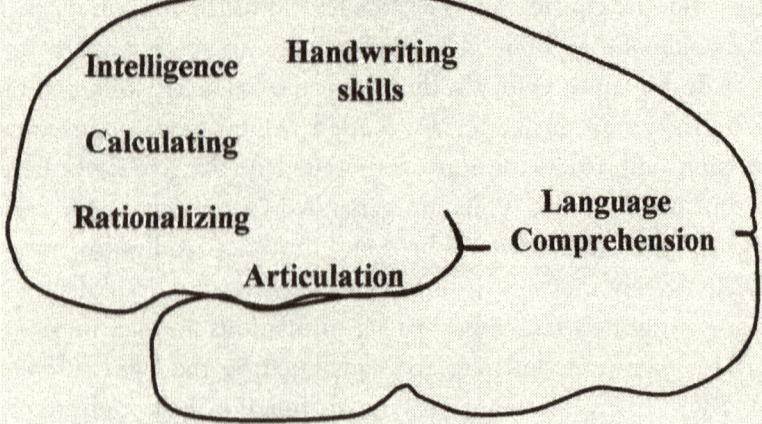

Figure 30. Differences in higher level functions of the right and left hemispheres of the brain (left side dominant individual).

Common Structures of the Human Brain and Their Function

Amygdaloid body: Located near the center of the brain, assists with learning and the processing of emotions.

Brain Stem: Located at the bottom of the brain, facilitates the transfer of information from the peripheral body up to the brain, and command instructions from the brain down to the body.

Broca area: Forward section of the brain in the inferior (lower) portion of the frontal lobe on the dominant side of the brain where speech is created. Commands are sent to the parietal lobe where speech is designed and muscle motor commands are generated.

Cerebellum: Area of the brain that interprets balance and position of the body, and position of the individual body parts.

Cerebral cortex: The upper portion of the brain, comprised of two frontal lobes, two parietal lobes, two occipital lobes, and two temporal lobes, which collectively engaged in higher level brain function and thought processes.

Cochlea: Rests inside, behind the right and left eardrums and changes acoustic waves into nerve impulses.

Diencephalon: Processes the data input from the brain stem and routes commands down from the cerebellum to the spinal cord.

Ears: Channel sound input to the eardrums.

Eardrum: Thin membrane located at the end of the ear canal, that resonates when sound strikes the membrane. Three interconnected bones transfer sound waves from the eardrum to the cochlea.

Eighth nerve nuclei (right and left): accept the nerve impulses from the cochlea and transmit them to Wernicke's area in the temporal lobe where the sound data is interpreted, speech created by another individual is interpreted.

Esophagus: tube-like passage way between the mouth and stomach that allows food to pass from the mouth to the stomach.

Eyes: accepts light input, transmitting the light to the retina at the rear of the eyeball, which converts the light signals to nerve impulses.

Frontal lobes: Personality, generation of ideas. The consciousness. Inferior portion on the dominant side of the brain is where speech is generated.

Hippocampus: curved band located in the temporal lobe, responsible fore memory and processing of new information. The subiculum, a division of the hippocampus, helps recognize danger and reward.

Occipital lobes: where the visual data captured by the eyes is interpreted.

Olfactory sensors: Sensors in the nose that sense smell.

Parietal lobes: design the strategy and execution of motor function.

Pituitary gland: origin of a number of neuroendocrine hormones.

Pineal gland: origin of the production of melatonin.

Retina: nerve that takes light accepted by the eyes and changes the light signals to nerve impulses then sends the data along the optic nerves (right and left) to the occipital lobes located at the rear of the brain. Some optic nerve fibers travel to the pineal gland and the SCN.

Suprachiasmatic nucleus (SCN): Acts as body's internal clock, setting time on a 25 hour cycle called the circadian rhythm. Light input and knowing time from a clock resets the circadian rhythm to a twenty-four hour solar cycle.

Temporal lobes: Stores long-term memory. Music interpretation.

Thalamus: located above the midbrain and brainstem. Acts as part of the human brain's central processor, a component of the diencephalon. The subconscious.

Tongue: Assists in moving food through the mouth to the esophagus, and has taste buds that act to sense the taste of food.

Ventral striatum: Located between the thalamus and the prefrontal cortex helps the brain process rewards.

Vestibular apparatus: Three loops located near the cochlea that provide information of general balance for the body.

The complex structures of the human brain and human body start to look a lot like components of the average desktop computer, when associations between the two are drawn. The following list begins to make some comparisons between the human brain and body, and the desktop computer:

Human Computer compared to Desktop Computer
Individual Components

Broca area=speech generation software
Cerebellum=position/balance interpretation card/software
Cochlea=Analog-to-Digital Converter in a sound card
Diencephalon=Part of the Human Central Processing Unit
Ears=Microphone
Eye=Optical camera
Frontal lobes=Memory: Operating system, Scratch pad memory, RAM, Audio card.
Lentiform nucleus=Part of the Human Central Processing Unit
Mouth/Vocal cords=speaker
Occipital lobes=Video card
Olfactory sensors=air sniffing sensor
Optic nerve=Coaxial cable.
Parietal lobes=Printer function, speaker output card (Digital-to-Analog Converter)
Suprachiasmatic nucleus (SCN)=Internal Time Clock
Temperature receptor=Heat probe
Temporal lobes=Memory: Long Term Storage. Music interpretation
Tongue=taste sensing probe
Touch receptor=Pressure sensor
Vestibular apparatus (in the ears)=gyroscope
Wernicke's area=speech recognition software

Flipping the descriptions around, common terms associated with components of the desktop computer can be associated with parts of the human brain and human body. Some of these comparisons are as follows:

Desktop Computer compared to the Human Computer Individual Components

Audio card=inferior frontal lobe on the dominant side
Analog-to-Digital converter=cochlea
CPU=diencephalon & lentiform nucleus
Coaxial cable=Optic nerve
Digital-to-Analog converter=Voice box
Floppy disc drive=Written word or recorded voice
Hard drive=Frontal and Temporal lobes
Internal clock=Suprachiasmatic nucleus (Hypothalamus)
Optical camera=Eyes
Power supply=Heart (at the cellular level: mitochondria)
Printer=Hands & a writing instrument
Printer card=Parietal lobes
Random Access Memory (RAM)=Frontal lobes
Read Only Memory (ROM)=Instinct or Preset memory files
Sound card=Wernicke's area
Speakers=Mouth & vocal cords
Video card=Occipital lobes
Video output card=Parietal lobes
Video screen output=Muscle motor function in the face, extremities

MEMORY

Read Only Memory (ROM) memory files are instinctual or preset memory files. Short-term memory is stored in the front portion of the brain in the frontal lobes where our personality exists. Long-term memory is stored in the back portion of the temporal lobes of the brain. Having long term memory in the back portion of the brain places it near the occipital lobes where sight is interpreted, and where speech is formulated.

Sleep, when we reach deep sleep or REM (rapid eye movement sleep), is thought to be where the short-term memory transfers information to long-term memory storage. During REM sleep, the short-term memory may clear its memory files, discarding meaningless data, and storing important data that the brain feels is required or important to retain.

DATA INPUT

The desktop computer receives input from the computer's keyboard, peripheral disk drives, scanner, optical camera, infrared port, the modem, etc. (see Figure 24).

The human brain receives input from all four of the body's limbs, and all of the internal organs (see Figure 31). These peripheral devices send their data inputs up the spinal cord to the brain stem. Also at the level of the brainstem, the medulla, pons and midbrain, facial nerves and the special senses send their data input to nerve nuclei located in the brainstem. Located above midbrain is the diencephalon. The midbrain coordinates the nerve inputs from all of the lower nerve tracts into the human CPU, comprised of the diencephalon, the lentiform nucleus, the amygdala nucleus, the caudate nucleus, and the medial and lateral geniculate nuclei.

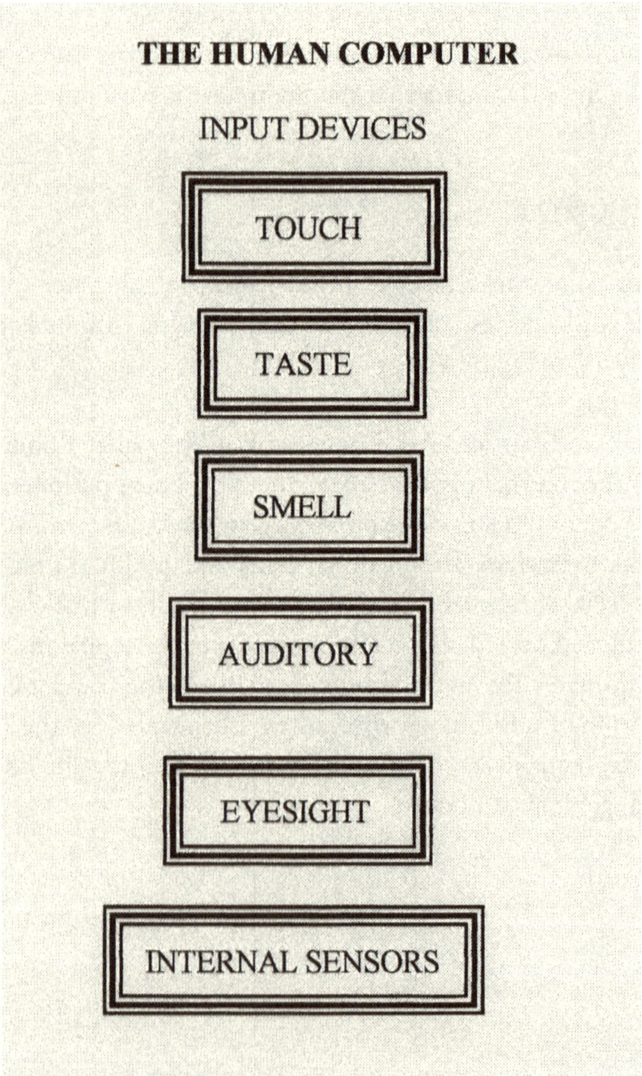

THE HUMAN COMPUTER

INPUT DEVICES

TOUCH

TASTE

SMELL

AUDITORY

EYESIGHT

INTERNAL SENSORS

Figure 31. Inputs to the Human Brain

DATA OUTPUT

The desktop computer's CPU commands the peripheral devices connected to the computer. The CPU sends instructions and routes data to the memory devices or peripheral devices such as the printer, the computer screen, infrared port, the modem, or peripheral disk drives such as floppy disc drives, CD drives or peripheral hard drives (see Figures 25 and 27).

The human computer's CPU at the level of diencephalon routes commands to the remainder of the body (see Figure 32). Commands may travel down to the level of the brainstem, and be sent out to facial or head and neck muscles, by way of facial nerves connected to the facial nerve nuclei located in the brainstem. Other commands destined for the body other than the face, may be directed downward through the spinal cord.

As the nerve impulses travel down the spinal cord, at the lower level of the brainstem, the nerves switch side. Impulses from the left side of the brain control the right side of the body, the right side of the brain controls the left side of the body. Peripheral nerves take impulses from the spinal nerve tracts to the arms, chest, abdomen and legs.

The human brain is comprised of the conscious, subconscious, long term memory, short-term memory, archived memory, language skills, scratch pad memory, and the logic unit (the interrupt controller). See Figure 33.

Supporting this core unit of the human brain are the input devices, output devices, the input-output devices including the emotional and sex units, and the imagination generation unit. See Figure 34.

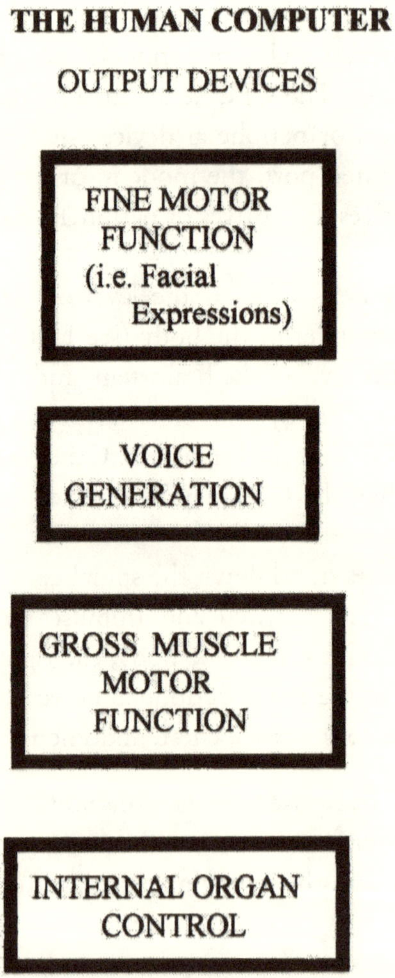

THE HUMAN COMPUTER

OUTPUT DEVICES

FINE MOTOR
FUNCTION
(i.e. Facial
        Expressions)

VOICE
GENERATION

GROSS MUSCLE
MOTOR
FUNCTION

INTERNAL ORGAN
CONTROL

Figure 32. Outputs of the Human Brain

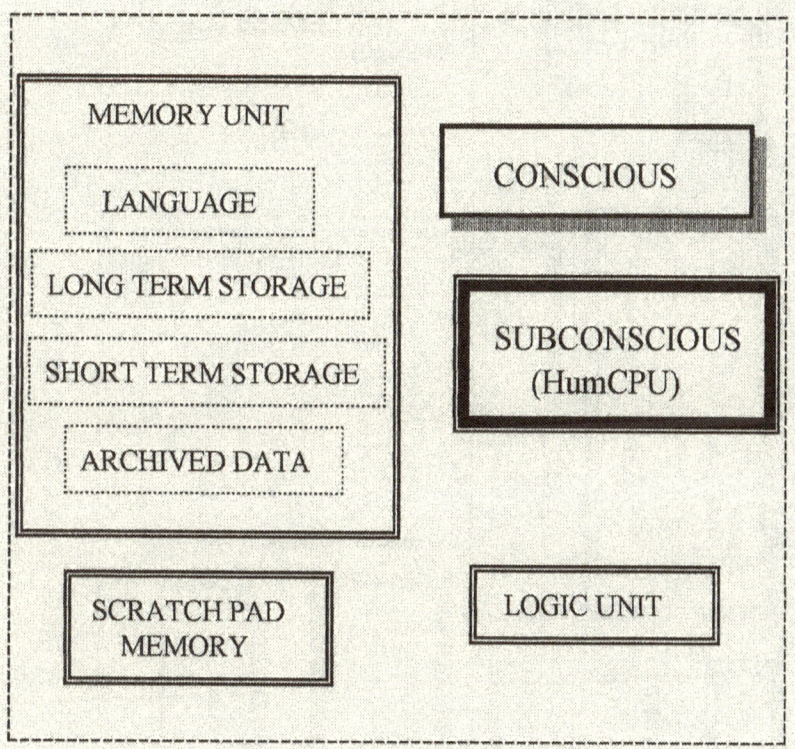

THE HUMAN COMPUTER

THE CORE DEVICE

MEMORY UNIT

LANGUAGE

LONG TERM STORAGE

SHORT TERM STORAGE

ARCHIVED DATA

CONSCIOUS

SUBCONSCIOUS
(HumCPU)

SCRATCH PAD
MEMORY

LOGIC UNIT

Figure 33. Consciousness, Hum CPU, and
Memory allocations

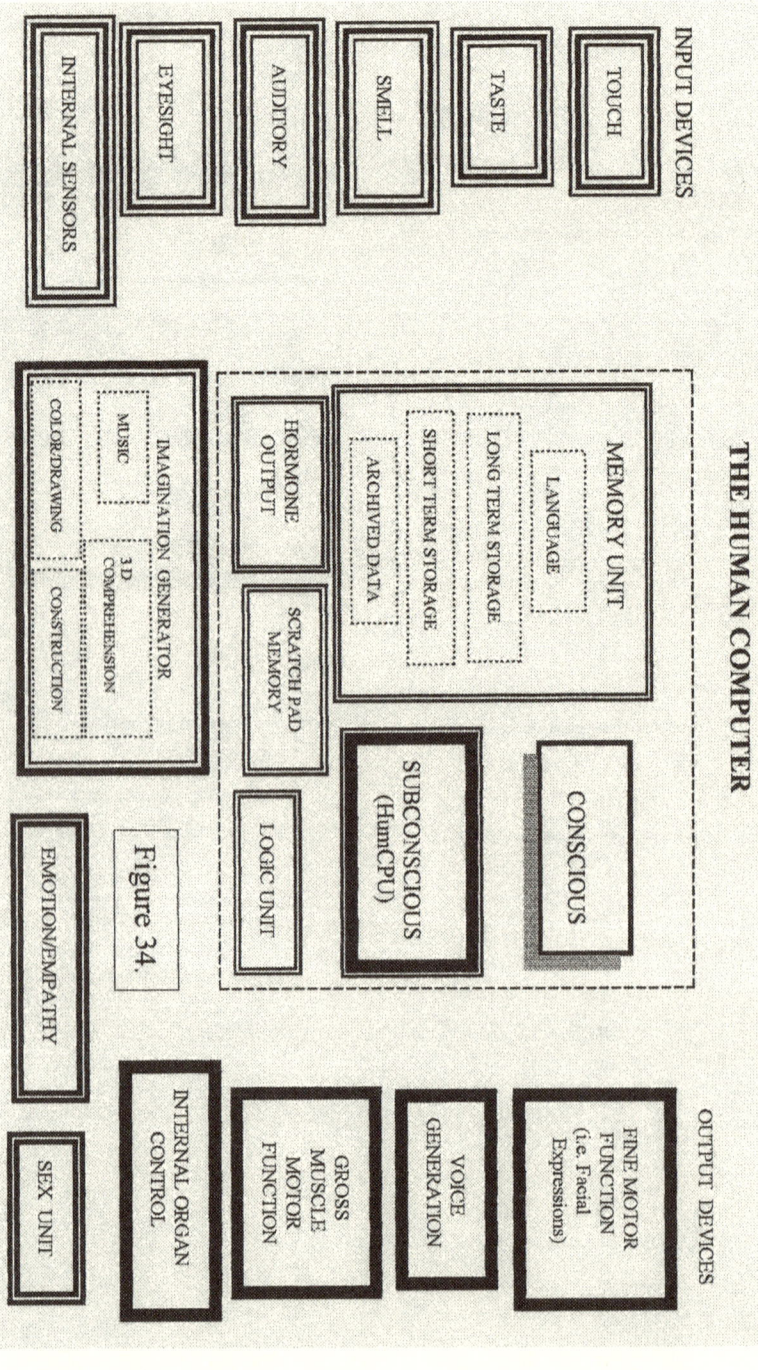

Figure 34.

Making comparisons between Figure 27 and Figure 34, a number of astonishing similarities exist. Both the desktop and biologic computer have input devices. Both store and retrieve information collected by the input devices. Both process huge amounts of information. Both rely on an internal clock. Both rely on an operating system that must be up-loaded into the system before the system can be interactive. Both engage in multi-tasking and use a central processor and co-processors. Both convert analog data to digital data, and digital to analog. Both have output devices that include speakers and motor functions.

The Human computer is required to engage in a number of complex physiologic tasks that the average desktop computer is not presently designed to carry out. The Human computer, therefore, is a more sophisticated computing device...for the present. As time marches on, and human ingenuity continues to produce faster computers with greater memory devices, and more efficient multi-tasking functions, the two...the desktop computer and...the biologic computer, in our heads,...are edging closer and closer together, and looking very much alike.

16

LOGIC, REASON, EMBARRASSMENT,... HOW IMPORTANT ARE <u>YOUR</u> INTERRUPTS?

As discussed earlier, an *interrupt* is a signal to the computer processor to stop what it is doing and attend to a new task that carries a higher priority than the current task. A desktop CPU may be faced with numerous interrupt commands at any given moment. Each interrupt has a priority assigned to it. The interrupt with the highest priority, is attended to first. As an interrupt command is addressed and the task completed, the desktop CPU then addresses any remaining interrupt commands again in the order of their priority designation, the highest priority first.

Humans are continuously faced with prioritizing interrupts both at the conscious and subconscious levels. Most humans possess a limited amount of money. Therefore, most humans find themselves, on a daily basis, prioritizing how to allocate their finances. Needs very commonly will create interrupt commands that cause us to spend money on necessary items rather than spending our money on leisure items. At a subconscious level, the body is always prioritizing its need for sleep versus a need to eat versus a need to be warmer or cooler versus a need to breath, a need to take in fluids, versus a need to locate a sexual partner, versus a number of other physiologic necessities the body has. The human conscious and the human subconscious are continuously attending to interrupts depending upon the priority they are assigned at any given moment in time (see Figures 35 and 36). Prioritizing needs helps the system run smoothly.

A. Logic: Is a number of interrupts

Logic, as it turns out, is merely a series of interrupts that help coordinate practical function of the human body. There are four major interrupt categories. These categories are (1) Learned interrupts, (2) Internal fear interrupts, otherwise referred to as instinct (also called phobias when the interrupt signal is too strong), (3) Physical interrupts, and (4) Emotional interrupts.

Learned interrupts can be social rules that we are taught or may be derived from experience. An example of a social interrupt then is halting at a stop sign. A learned interrupt may be not sticking your hand in a fire after you burned your finger as a kid. An instinct, is not burning your finger in the fire as a child or as an adult because there is already an interrupt with a high enough priority to prevent you from doing that. A phobia is that you are deathly afraid of fire because an inherent interrupt in you has a higher priority than normal.

Internal fears are interrupts we are born with. When these inborn interrupts are excessive in their expression, they may be referred to as 'phobias'. But, in moderation, internal fears help in survival of the individual. Internal fears warn us of the danger of heights, danger of fire, dangers that may be lurking in shadows or in the dark in general.

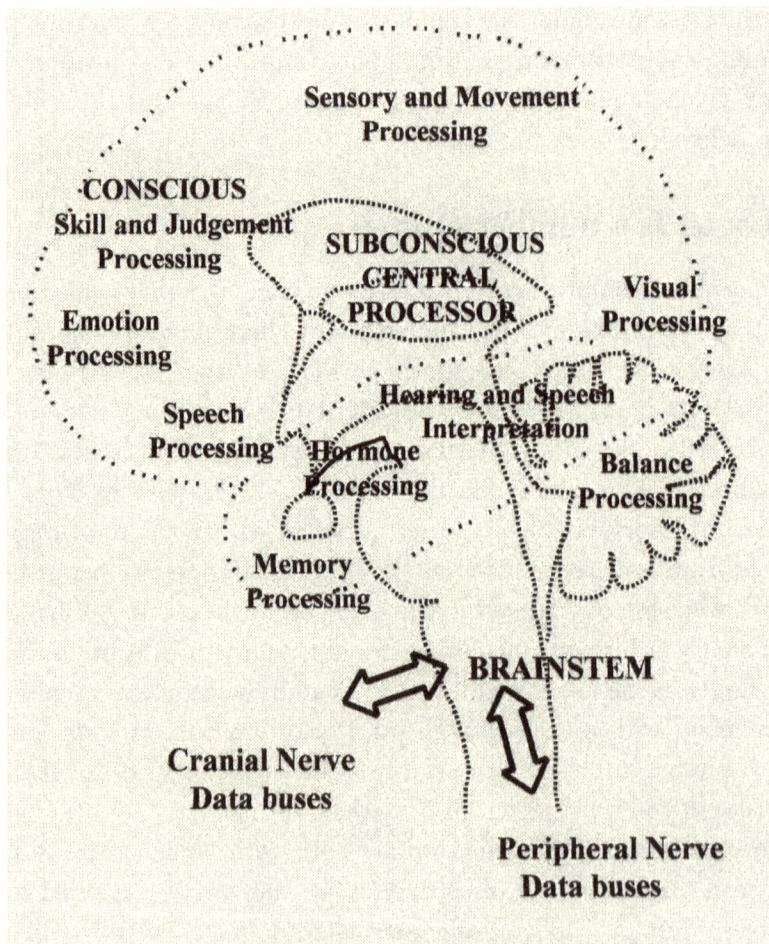

Figure 35. Different processing
subdivisions of the brain.

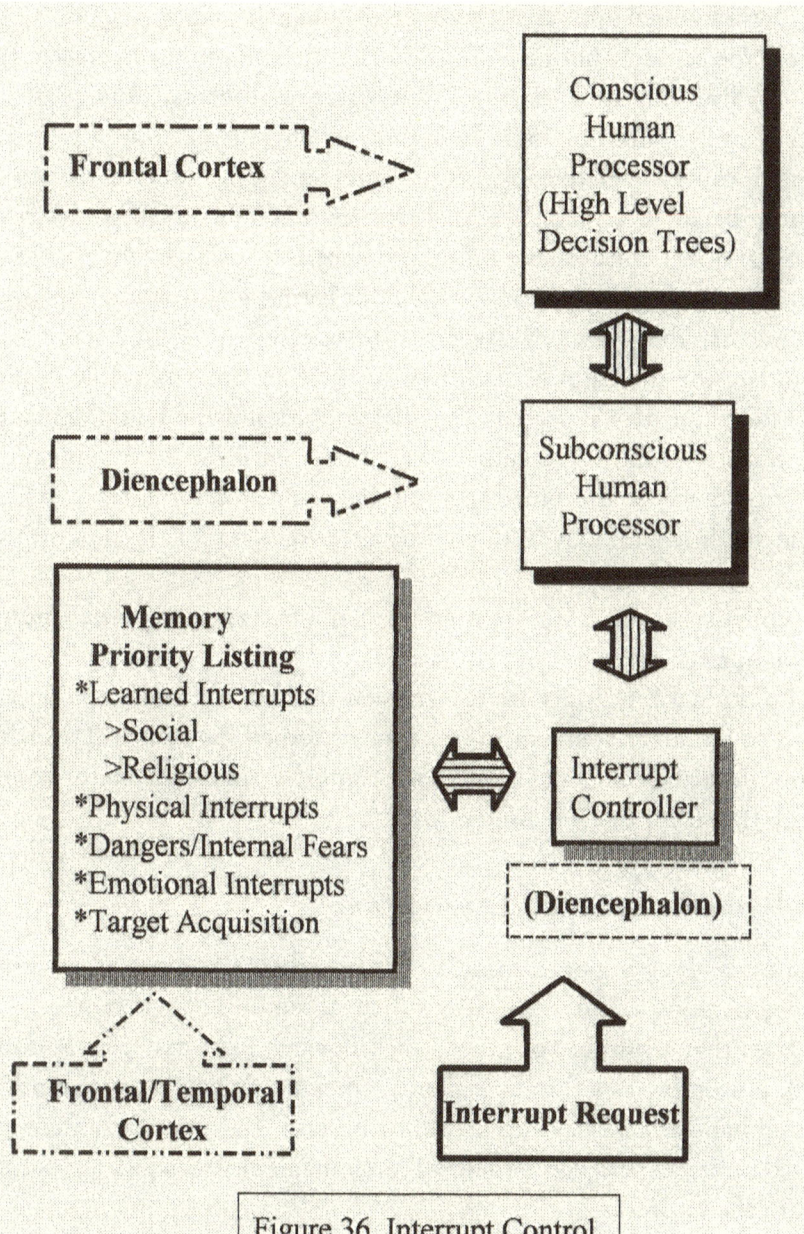

Figure 36. Interrupt Control

Physical interrupts are the sense of adequate air intake, sense of thirst, the sense of hunger. We can live only five minutes without air, we can live only three days without water, and we can survive up to three weeks without food. Therefore, when our air supply is interrupted, our body activates a strong interrupt to all other activities to search for air if needed. We may temper this interrupt if we learn to swim and feel comfortable holding our breath for a certain period of time. But if the time period we hold our breath goes too long, or if we realize that the source of oxygen is outside our reach, the body may reinitiate the interrupt as the physical need grows in importance and this interrupt may supercede all other activities of the body, demanding full attention of all faculties of the body, until the need to breath is satisfied, and we are breathing normally and comfortable, at which time, the interrupt may be down regulated to its normal level of importance.

Thirst and hunger can become very important interrupts depending upon our last drink or our last meal.

Emotional interrupts are driven by a myriad of human emotion. I love, or feel an attachment to someone or something, so I will override other interrupts in order to do something for this someone or something that I feel the attachment for.

B: Reason: Learned Interrupts

Reason is observing or learning, or experiencing, a number of processes and generating an interrupt that is then stored in the logic center.

I see that a number of people pick flowers. I see that people don't pick poison ivy as a flower. I learn by picking poison ivy with unprotected hands that I develop an annoying rash. I generate an interrupt, based on my experience, which will discourage me from picking poison ivy in the future.

C: Ambition: Who we are meant to be

Ambition is a center that works directly with the conscious mental processor. Ambition is defined as, who we are, and what we think we should be, in other words 'what is self'. It is also the blending of the fight or flight centers in the brain. As we may recall when grossly presented with a threat, our brain will respond by trying to decipher whether to mount a response against the threat, which most of us would interpret as 'fighting', or flee the threat, which most of us would interpret as 'flight'.

Ambition is a series of subroutines that define what is the individual's self and self-worth. The Ambition center stimulates the conscious mind, which in turn stimulates the subconscious processor to effect actions to reach goals that help define what self is. Goals may be learned. Goals may be retrieved from the archived memory. A person may desire to be a police officer, because archived memory files send the template (series of subroutines that define being a law enforcement officer) stimulate the ambition center to set being a police officer as a goal in life. Reaching the operating level of being a police office may in turn then stimulate the pleasure center in the body which keeps the individual conducting himself or herself as a police office. Other templates may exist. There may be a template for being a store owner, a chef, a farmer, a butcher, an architect, a construction worker, a judge. Having templates released from archived memory into the ambition centers of individuals ensures that a social structure exists in communities.

To pursue a goal involves generating a fight response that is supported by the pleasure center in the brain. Pursuing and achieving significant goals in our lives often requires <u>overcoming</u> various interrupt controls, and paradoxically makes us feel good.

D. Empathy Center: Catch the Wave

Empathy is being able to understand and relate to the experiences or actions of other humans. Empathy helps to build and define a spirit of 'community' amongst humans. Empathy helps us recognize when other humans are struggling, and sensitizes us to take action to help the struggling human. Emotions are generated by the empathy center. Emotions are waves of action given a high priority interrupt which is the result of some valuable experience or event.

Emotions interrupt the logic center with a higher priority interrupt command causing some action to be considered or to be effected by the subconscious. Emotions also up-regulate or down-regulate the pleasure center causing the conscious mind to become enamored by the interrupt and cause the conscious center to stimulate the subconscious processor to act on the high priority interrupt caused by the emotion. More than one emotion occurring at one time represents more than one high priority interrupt attempting to command the subconscious and therefore results in a state of confusion by the conscious mind processor and disorder in the subconscious mind processor.

E. Embarrassment

Embarrassment is a devastating emotion. Embarrassment is generated by an occurrence involving an individual that threatens the individual's self worth. When such an occurrence happens to the point that the individual's self worth is compromised, at the level of the Ambition center, where the fight and flight centers are located, the flight interrupt subroutines are activated, and a person is driven to detach themselves from the situation. Embarrassment is likened to the activation of a hurricane type blast interrupt.

F. Pain

Pain can, at times, become the most powerful interrupt signal generated by the body. Chronic pain can be a devastating, possibly disabling interrupt signal in the body.

Pain is best described as a sensor in the skin, or a sensor embedded in an internal organ, when stimulated, sends a distress signal to the brain. The aim of the distress signal is to alert the brain that some harm has occurred to the affected portion of a body. The meaning of the distress signal is to alert the brain to relocate the affected portion of the body out of harm's way, and/or to care for or seek care of the affected portion of the body. Pain signals may result from tissue damage due to trauma or the invasion of a pathogen such as a bacterial infection or the invasion of tumor cells upon normal tissue cells, or tissue damage due to a toxic chemical, a medication in higher than normal dosages, or in the case of the autoimmune diseases an individual's own immune system may attack and destroy healthy tissue.

Pain may be beneficial in saving a limb from injury. When a hand or a foot is suddenly exposed to the flames of a fire, reflexively the exposed limb is often removed from danger before the mind registers pain. That is, the subconscious receives the pain signal and acts to remove the affected limb from danger, before the conscious mind has recognized the danger or felt the pain. If tissue damage has occurred in the affected limb, then pain may occur, but usually after the limb has been rescued from the immediate danger.

In the face of immediate tissue damage, pain acts as the most influential interrupt in the body causing the body to react to the pain by changing limb position or body location or placing the body in a defensive or aggressive posture depending upon the nature of the threat.

In a situation where tissue damage may be occurring, but at a slow rate, such as what might result from the growth of cancerous cells invading normal tissue, the pain sensor may react to the damage setting off a constant, unremitting pain signal to the brain. This signal may cause the human brain to seek medical help to try to treat such a cancer. In the case of a desktop computer connected to a series of sensors, the sensors may produce nuisance signals. Such nuisance signals can be filtered out. In the human brain, if tissue damage is occurring, such signals, even if constant, are not filtered out, otherwise ongoing tissue damage would be ignored.

Often times, injury to the human body will result in the generation of pain. Such pain interrupts our normal daily routine causing the injured individual to either rest the injured portion of the body or seek medical attention. If the cause of the injury is known, and the damage is not too severe, healing will most likely occur and the pain signal generated by the damaged tissue will diminish as the healing process completes itself. If tissue injury occurs without a clear cause and effect relationship, or if the injury is severe, the pain signal may not diminish quickly. The subconscious may communicate with the conscious level quite often. The conscious mind may find itself interrupted by the persistent pain signal generated by the injured tissue. A chronic pain cycle may be generated. The concerned conscious mind may periodically stop what it is doing and ask the subconscious mind if the affected part of the body is still sending out a pain signal. In the case of no apparent cause for the onset of the pain, the conscious mind may be very interested in the progress of the pain signal. The upper cortical functions of the brain may become tied up in concern over the presence of the pain signal.

The role of a physician is often to identify the etiology of the tissue damage whether it be related to an infection, a disease, a tumor, an allergic reaction by the body or a toxic chemical. In the case of an injury, the physician is often skilled at determining the extent of the tissue damage and the cause and effect relationship. Simply understanding the physical reason for the cause of one's pain, and understanding the extent of the injury, often time provides a patient's conscious and subconscious levels with enough information to be able to down-regulate a pain signal. Often in the cases of chronic pain, physicians will introduce pain medications whether they be nonsteroidal anti-inflammatory drugs called NSAIDs, or simply analgesic medications, such as acetaminophen, or stronger pain medications often referred to as narcotics. Adding a pain medication, often has the result of disabling the chronic pain signal by blocking pain receptors. Once the pain signal has been disrupted over a sufficient period of time, this signal is often down-regulated as long as ongoing tissue damage is not continuing to occur.

Interrupts are extremely important in making sure that human body recognizes tissue injury, tissue death, or infection. The assigned priority level of interrupts creates a rank and order that results in a smooth flow of decisions and actions for the body.

When Interrupts Are Too Much

Too many interrupts and we start feel overwhelmed. In a computer, if the interrupt system fails to conduct itself in an orderly manner, we all know this to result in the computer locking up, the computer screen freezes, unable to respond to commands input from devices such as the keyboard or the mouse. In a human, if the interrupt system fails to execute in an orderly manner, and conflicting interrupts occupy our brain, then confusion and indecision ensue. When the level of confusion stretches beyond the level of processing power, a brain freeze ensues, a nervous breakdown occurs; medical attention may be required. In the worst case scenario, a brain infarct (stroke) might occur, resulting in permanent tissue damage.

On the other hand, interrupts, in general, are vitally important to assure a timely and organized manner of normal brain function.

So let's boil it down. Those considered to be overly *logical* people, such as engineers, generally have very strong *learned* interrupts. Those that are considered overly empathetic or emotional, generally have very strong *emotional* interrupts. Those that are social geeks, have very strong *social* interrupts. Those that are unable to manage their interrupts effectively, or efficiently might be considered confused, possibly even a *little* crazy.

17

IS YOUR BOSS ON THE JOB? SOFTWARE FOR THE BRAIN

If your personal desktop computer's hard drive experiences a nonfatal error, one might try to use a utility program to access the rudimentary Disc Operating System (DOS) program to try to activate DOS features that will assist in fixing the error, in order to bring the hard drive back on line. The DOS is responsible for organization and maintenance of the computer programs on file in the computer's memory. Present Windows software, which is much more user-friendly to the average computer operator, is built on top of the DOS system programming.

The DOS program acts as the fundamental program instructions for controlling the activities of the hard drive. The question becomes, does the human brain have a similar controlling program? Such a program could be termed the <u>B</u>rain <u>O</u>perating <u>S</u>ystem <u>S</u>oftware (BOSS). A program such as the BOSS may be the operating system that controls how the subconscious brain functions.

We think in terms of accessing the human brain by communicating with a person's conscious brain, either through speech, sign language, pictures or other stimuli. We are capable of only crudely studying brain function by monitoring electrical wave patterns emitted by the brain using an electroencephalogram (EEG) device. We can check a person's evoked potentials that give physicians a crude sense of the correlation between light reflexes and brain function.

What if there is an underlying BOSS in the human brain? There may be a means to access the human brain by learning the underlying software codes that the human brain uses. Understanding the BOSS may allow us to access the subconscious brain and its processing power directly. We may be able to discover BOSS commands that allows future generations of physicians to communicate directly with the brain's operating system and search out failures in the brain. This could become especially helpful in treating patients that are in a coma and unresponsive in the conventional sense. There may even be a means of writing biologic program patches that might fix malfunctions that occur in the brain's operating system software or other neurologic deficits.

Curing Dementia and Alzheimer's disease

For ages, physicians have thought **disease** to be a '**structural**' process or problem.

This structural concept for disease, till now, has been acceptable. But regarding every disease state in the human body to be structural in origin would be analogous to saying that all computer problems are solely hardware problems in origin; which we certainly, those that work with computers, know <u>not</u> to be true.

Actually, we already know that not all disease states are hardware or structural in origin. We know that viruses (many of which are responsible for the common cold or the flu or HIV (the virus responsible for the AIDS epidemic)) attack normal cells, invade the cell's nucleus and insert their viral DNA material into the normal cell's DNA. By the virus inserting its instruction code into the normal cell's instruction code, the virus enslaves the normal cell. The virus's instruction code causes the normal cell to cease its own cellular functions and forces the normal cell to use its resources to produce copies of the virus. Once the copies of the virus have become too overwhelming in numbers, the normal cell ruptures, thus releasing the copies of the virus so that they can go on to invade other normal cells. This scenario depicts a very common example where a software alteration results in physical changes to normal cells and thus a state of disease. (Actually, this is very similar to how computer hackers compose computer viruses to invade target computers.)

The present definition of disease includes an abnormal anatomy or function of an organ or system of an organism, resulting from various causes, including infection, genetic defect, environmental stress, and characterized by an identifiable group of signs or symptoms. This classic definition may need to be broadened to include alterations in brain operating software may lead to physical changes in the body, or functional changes in the central or peripheral nervous systems.

To take the concept of brain software a step forward, the conditions of dementia (the deterioration of mental faculties such as memory, concentration, and judgment, resulting from organic disease or a disorder of the brain) and Alzheimer's disease (the decrease in brain function due to a degeneration of brain cells and infiltration of abnormal proteins along nerve cells which interfere with nerve signal transmission) may be treated with brain software reboots, upgrades, or software patches.

When attempting to protect our working software on a desktop computer from hard drive crashes, we prepare recovery discs so that in the event the computer fails, it can be rebooted. Understanding the brain operation software system may evolve to the point someday, we may not only be able to download software patches, but we may be able to make a copy of a person's brain operating system software (BOSS) and store it, so that if a person's brain failed, became demented or developed Alzheimer's disease, the person's brain operating system could be rebooted into his or her brain.

Multiple personalities

We know in some people, there have been reported multiple personalities. Possibly, in some of these people, more than one operating system is functioning at any given time. An alternative explanation would be that people experiencing multiple personalities carry separate, noncommunicating memory files. We know in the desktop computer that we can partition the hard drive into two or more independent sections. Possibly, a human brain's memory can also undergo a form or partitioning whereby one part of the memory does not communicate with other parts; therefore, stored memories may end up in two or more partitions and appear to be separate and distinct collections of experiences.

Pioneering a new day in medicine

By today's standard, if an individual suffers a heart attack and collapses to the floor, EMS is alerted. If one survives till an emergency technician reaches the victim's side, the person will be rushed to the closest emergency facility. Upon bursting through the doors of the ER, the emergency room physician will order that the blood pressure and pulse will be regularly monitored. The physician will have an EKG device attached to the patient's chest to monitor the heart rhythm. Blood will be drawn to search for enzymes that might have been released into the bloodstream that would indicate damage to heart tissue due to ischemia. Even if the person's condition is fully stabilized, such that the patient recovers and feels well, if the blood tests suggest sufficient damage to the heart muscle has occurred (considered a heart attack), the patient is hospitalized.

What if instead of checking and monitoring external signs and symptoms, the technology were sophisticated enough to tap into the data the brain was privileged to?

Let's say when the above mentioned heart attack victim rolled through the doors of the local ER, in addition to hooking up a blood pressure and pulse monitor, and attaching an EKG machine to the patient's chest, the ER physician was able to directly tap into and communicate with the individual's brain. It would make sense that the brain would know full well what was happening to the heart, and how severe the damage was that had occurred. The brain is most likely privileged to blood pressure, pulse, temperature and other important physiologic data. Possibly all that we need to know about any disease state, the brain could communicate vital information to physicians if only we knew how to communicated directly with the brain. Remember Bones from Star Trek, waving a sensor probe across a person's chest? If a sensor probe could communicate directly with the brain's operating software system, possibly vital information could be downloaded to a remote sensor, analyzed, and then healing commands could be uploaded to the brain, to effect repairs to injuries or cure diseases.

The more we learn about how the human brain functions, the more we will be able to turn this knowledge into tremendous advances in medical science and computer technology.

18

BOOTING UP THE HUMAN COMPUTER

Unlike the desktop computer, booting up the human computer must be considered in two related, but unique manners. Where a store bought desktop computer comes with many faculties ready for use, the human brain must undergo significant maturation and development of the brain cells, before it is ready to be used. The desktop computer might arrive at one's house or business all set and ready to go with pre-loaded software, or one might need to install software programs in order to facilitate the computer to run the way the user would like it to run. Many computer users will add such things as word processing software, money management programs, spread sheet software, photo processing software, game programs, and screen saver software, to customize their computers to their interests. Like the desktop computer, the human brain must be loaded with useful software. This software, or brainware, is loaded during the early pre-school learning phase, and the lengthy educational process, which customarily involves, kindergarten, elementary school, middle school, high school, college, and for some, graduate school (see Figure 37).

The Human Brain Grows Up

Like the remainder of the body, the brain requires years to develop to the adult capacity. The prefrontal cortex, the most forward portion of the cerebral cortex, appears to develop last. The prefrontal cortex is believed to be where humans control their emotions and where the human develops strategy and planning. Since the prefrontal cortex develops rather late, it may account for the emotional outbursts and the lack of clear planning and judgment some teenagers are known to exhibit. I remember going through my late teens and early twenties feeling like I couldn't walk straight or chew gum right. I wanted to do everything right. I wanted to act with the suave and confidence of a forty-five year old, while being twenty-one. I felt like the mouth, brain, arms and legs just did not act in an orchestrated manner. For while there, I felt like an absolute klutz.

Personal Computer

Division of the Hardrive

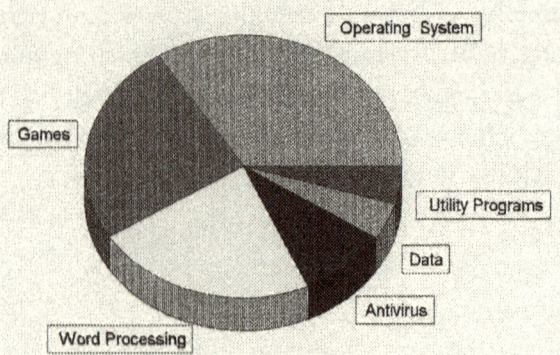

Human Brain

Division of the Upper Cortical Memory

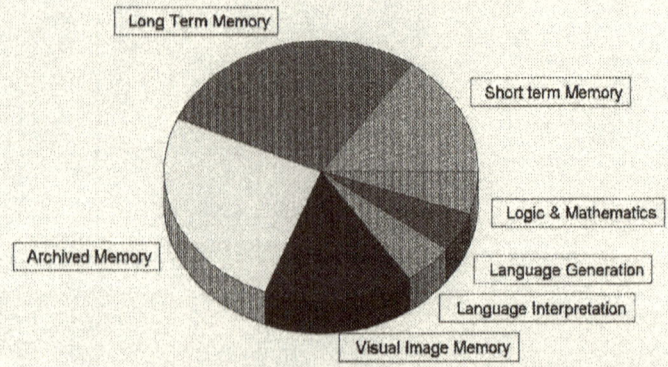

Figure 37. Typical organization of a computer's hard drive and possible organization of the human brain's memory files.

As the brain tissue develops, nerve synaptic routes develop. The human brain is somewhat adaptable to its surrounding environment. Depending upon what a young developing human is called upon to learn or called upon to do, nerve pathways may route differently. Unused or unlearned pathways may remain dormant and may not fully mature. The human brain may enhance its resources of brainpower in areas that it feels it may need to be successful or survive in the future.

An example may be an engineering student who nearly fails freshmen English in college. Later, working as a professional engineer, the individual may be called upon to engage in a significant amount of writing to produce work proposals, study designs and study reports. Though the individual is not strong in his or her writing skills, the duties of their job and career require, and challenge, the individual to develop writing skills or lose their job.

The Human Brain Gets Educated

Humans begin life dependant upon others for tending to, and fulfilling, all of their needs. Humans are capable of incorporating into any culture. Like a computer that is able to load a number of different operating systems, the human brain is able to load a number of different languages and use any one of these operating systems to communicate with other humans. Human languages are analogous to the communications portion of the operating system(s) in a computer.

Language allows humans to communicate with each other in a fluent, orderly fashion. The peripheral devices of the vocal cords that produce sounds, the muscle structures in our hands and face that produce body language, and facial expressions, facilitate our communicating with others. The peripheral devices of our ears that detect sounds, and our eyes that gather optical images of our surroundings, allow us to receive information from other humans as well as our surroundings. A complex subconscious operating system deciphers the incoming signals. Our conscious brain then responds to the incoming signals and produces outgoing communications signals. The ability to communicate with others facilitates the structure and network of society.

Developing a functional operating system, whether it be language in the form of spoken words, or written words or sign language, is paramount in one's ability to interact efficiently and effectively with other humans. It is generally recognized that the more articulate individuals are the more educated individuals in society, these individuals being either self-taught, extensive readers, and/or having attended a facility of higher education. One can become very articulate in their speech and/or their writing skills and/or their artistic skills with desire, training, and motivation.

Achievement of communication skills is linked to creating an operating system. Humans start out with no language skills other than rudimentary sounds and the ability to use their body movements to express their intentions. A child will often cry and use a finger to point at an object of desire, located outside their reach, when they do not possess an adequate vocabulary to formulate the words to communicate their intentions or desires. As a child grows, the vocabulary develops and speech becomes an integral part of daily life. The ability to articulate the speech is dependent upon the size of the vocabulary one acquires and stores in their brain. Acquiring large vocabularies is a function of exposing oneself to the words, their pronunciations and meanings and then storing these associated facts in the memory files of the brain.

Acquiring logic skills involves a similar process. Knowing how to construct objects such as drawings, puzzles, and toy buildings becomes a function of learning how individual pieces fit together. To build a play house, a child must learn how pieces can be set together so that walls fit properly and how to construct a roof on top of the foundation pieces. At higher levels, mathematical skills become important to acquire. Mathematics helps predict outcomes. Being able to calculate strength of objects allows one to predict how many pieces one needs to build a play house and allows one to predict if there are enough materials available to accomplish the planned task, and answers to questions such as 'will the planned foundation support the load of the roof', before the time and energy is spent starting the task.

Acquiring extensive logic skills allows an individual the ability to solve problems. Solving problems is an important component of any successful career. Usually, those individuals that can solve a companies problems, and in doing so, create means for a company to be productive and fruitful, are the employees that enjoy the profits of these same companies.

Most humans start out with the machinery to perform language and logic tasks. Most humans can learn almost any functional language and use it effectively. It is the art of learning language and logic that defines an individual's operating system. The more a person learns, the larger and the more articulate the operating system the individual's brain acquires.

Two equally capable individuals may experience two totally different lives. One may live in an impoverished environment and have little to no opportunity to acquire high levels of language or logic skills. This individual may lead a subservient and unproductive life. A similar individual may have the means of becoming educated, and the desire to be educated and go on to lead a fulfilling life in a successful career. On the other hand, the individual in the impoverished environment may have instilled in them a tremendous desire to learn that drives their future, while the individual that has all of the world's resources within reach, if they have no desire to learn, will fail to develop higher levels of language or learning skills.

Marie Curie, winner of two Nobel prizes, once commented, "*Chance favors the prepared mind.*" What she was identifying in her statement was that the more advanced an individual's learning, the more advanced the operating system the individual is then equipped with to analyze the facts they encounter to make new and important discoveries about the environment.

One would not expect an uneducated individual living and working in an inner-city ghetto to find a cure for breast cancer. It could happen, but without the proper resources and training, the chances are slim that if a computer file regarding a breast cancer cure were to flash in the individual's mind, that the individual would be able to act upon this invaluable event. This same individual, who believes in himself or herself, educates themselves, studies in medical school and later becomes a professor on staff at medical research institution, may indeed produce a cure for breast cancer. The only hope at arriving at such a cure is with acquiring, enhancing, and expanding the language and logic skills in the individual. To cure a disease such as breast cancer, requires the individual to develop the skills and depth of their operating system beyond that of their contemporaries, and sometimes in the process, it just requires blind luck…keep reading for further details….

But What is the Learning Process?

The learning process occurs in four major categories.

Category I:

We learn rules and regulations regarding our place in society, the expectations that society has of our behavior and to what extent we can expect how society may react given certain actions we may exhibit. We also learn the limits our body has, as well as the skills that we may be good at, whether such skills are physical or mental capacities.

Category II:

We learn image associations.

Recently with heightened security, we have been told of high-risk security points being monitored by automated surveillance equipment. The cameras at such a checkpoint, snap pictures of the faces of the people passing through the view range of the camera lens. The photographed image is digitized, and analyzed by facial landmark measurements. The pictures of the people are compared, in real-time, to a data bank of human pictures. If the computer finds a match, associated memory files are also retrieved including the person's name, what is known of the individual's profile, and the security risk the individual poses to the general public.

Our brain dissects the environment surrounding us into a data bank of unique objects. For each physical object we encounter, we assign a name to the object. Given the name of the object, we then assign a text description to the name. Therefore, in addition to the visual image file, the name file, and the text descriptive file that becomes stored in our memory, our brain also attaches several other files when appropriate. Almost all visual file images will also have a *threat assessment* file associated with it. When we see an object, we automatically determine the level of concern or harm the object might convey to us. In addition to threat assessment, other attachments include an emotional file, a taste file, a touch file, a smell file, and an audio file when appropriate and a list of associated files when appropriate.

We open our eyes, and we see a room around us. The room is filled with numerous objects. If the room is familiar to us, each object has a name; each object that we saw in the visual field is something that we could describe in words to another person, if asked. Each object would either evoke an emotion such as with personal items, or possibly evoke no emotion at all. A picture of one's mother fixed on the wall as a memento, might stir a variety of strong emotions inside our brain, while on the other hand the wall switch that turns the ceiling light on and off, may evoke no emotion whatsoever. We may remember the picture of our mother, and be able to name the picture and describe it and describe how it makes us feel, but we also, most likely, will remember where it is located by being able to recall other objects that are located near it, and the room that it is located in.

In addition, we may be driving down the street in a car (see Figure 38). When we see a car approaching us from the opposite direction, we recognize the image, place a name to the image, we attach a description to the name, we attach any associations to the image (such as if it is owned by a friend), we assign a threat to the image. If the car seems to be moving down the road at a proper speed and in an expected manner, the chances the car might hit you are low, the threat assessment is low; but if the car suddenly swerves in its lane, the threat or risk assessment may suddenly increase dramatically, and an increased attention, and or defensive action, may be required. Further, the image of the car may also have smell, touch and audio memory files attached to it. If the vehicle exhibits an unusual smell or sound, then the image may have a special olfactory or sound memory file assigned to it.

Category III:

We are able to learn abstract concepts. We are able to associate text names and text descriptions to objects that may not have a physical form or image. In addition to our opposable thumb (our thumb is able to rotate and make contact with the little finger), the fact that humans can think and remember abstract concepts, sets us apart from the rest of the animals in the animal kingdom that inhabit the Earth. The learning of basic mechanical skills has been referred to as *procedural* memory; where as the learning of new data based on information has been referred to as *declarative* memory, or has been described as '*knowing how*' versus '*knowing that*' respectively.

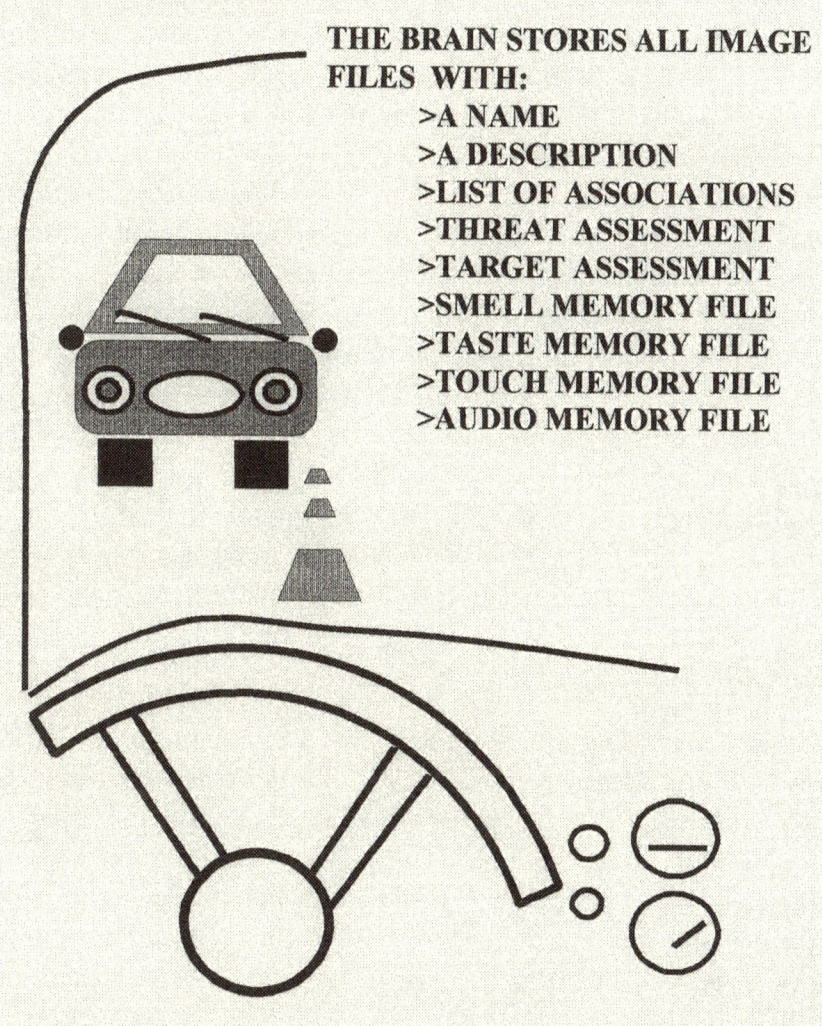

THE BRAIN STORES ALL IMAGE FILES WITH:
>A NAME
>A DESCRIPTION
>LIST OF ASSOCIATIONS
>THREAT ASSESSMENT
>TARGET ASSESSMENT
>SMELL MEMORY FILE
>TASTE MEMORY FILE
>TOUCH MEMORY FILE
>AUDIO MEMORY FILE

Figure 38. Driving down the road, spying a car driving in the opposite direction invokes an image file to be generated in the occipital lobes of the brain.

So humans are able to generate image files and apply text names and descriptions to these image files, even though our eyes may have never seen such an object or image. We are not only able to recognize objects and perform basic mechanical skills, but we are able to understand concepts.

I have never seen an active, erupting volcano. I have seen pictures of erupting volcanoes in books, and I have seen video clips of a volcano spewing forth lava in movies and on television. But, if I had never seen a picture of a volcano, and if someone who had seen a volcano described to me in words, what one looked like in its active form by saying 'a volcano is a large mountain with a crater at its summit, that periodically spews forth fire and thick black smoke up into the air, and belches out fiery, molten rock that pours down the sides of the mountain that consumes everything in its path,' my mind might be able to associate enough imagery from what files exist in my memory, to fabricate a mental image of what an erupting volcano might look like.

Category IV:

We are able to learn assimilated data files. The assimilation of experiences and knowledge results in the development of decision trees. The more knowledge and/or experience one can accumulate, the larger the decision trees can be developed. The larger the decision trees that one has stored in their brain, the more likely that an individual can analyze situations that they encounter and arrive at correct decisions in order to either achieve a positive outcome, or create the best chance of survival.

The brain uses all four major lobes to accomplish the learning process including the frontal, parietal, occipital and temporal lobes (see Figure 39). The various memory files are stored and utilized by different portions of the brain. <u>Frontal lobes</u> create decisional memory files (minute to minute, one has to make decisions and then be able to remember what decisions where made) and generates even more abstract forms of memory files such as creative thoughts and emotional reactions. The <u>parietal lobes</u> coordinate sensory input and body movement, therefore, require storing and utilizing various forms of data regarding the environment and where the body fits into the environment, and best means to maneuver around its surrounding environment. The <u>occipital lobes</u> produce visual memory files derived from what the eyes see. The <u>temporal lobes</u> coordinate the visual memories, the audio memories, the olfactory memories, and the tactile memories into long-term memory files.

The human learning process seems to be dependent upon those portions of the brain located deep inside the brain. Learning seems dependent upon a neurocircuit referred to as the Papez circuit. The Papez circuit appears to be comprised of neuropathways linking several internal brain structures together. The portions of the brain that appear to comprise the Papez circuit, and therefore important in the learning process, include the diencephalon, the lentiform nucleus, the hippocampus, the amygdala, the temporal lobes, and the hypothalamus. From the Papez circuit, information seems to be assimilated, then stored in the appropriate locations of the temporal lobes or other areas of the cerebral cortex.

Optimal booting up of the human computer requires a diligent, long and involved learning process. Achieving one's goals in life is dependent upon securing an optimal operating system and memory files. Education is the fundamental key for improving one's ultimate chance at being successful.

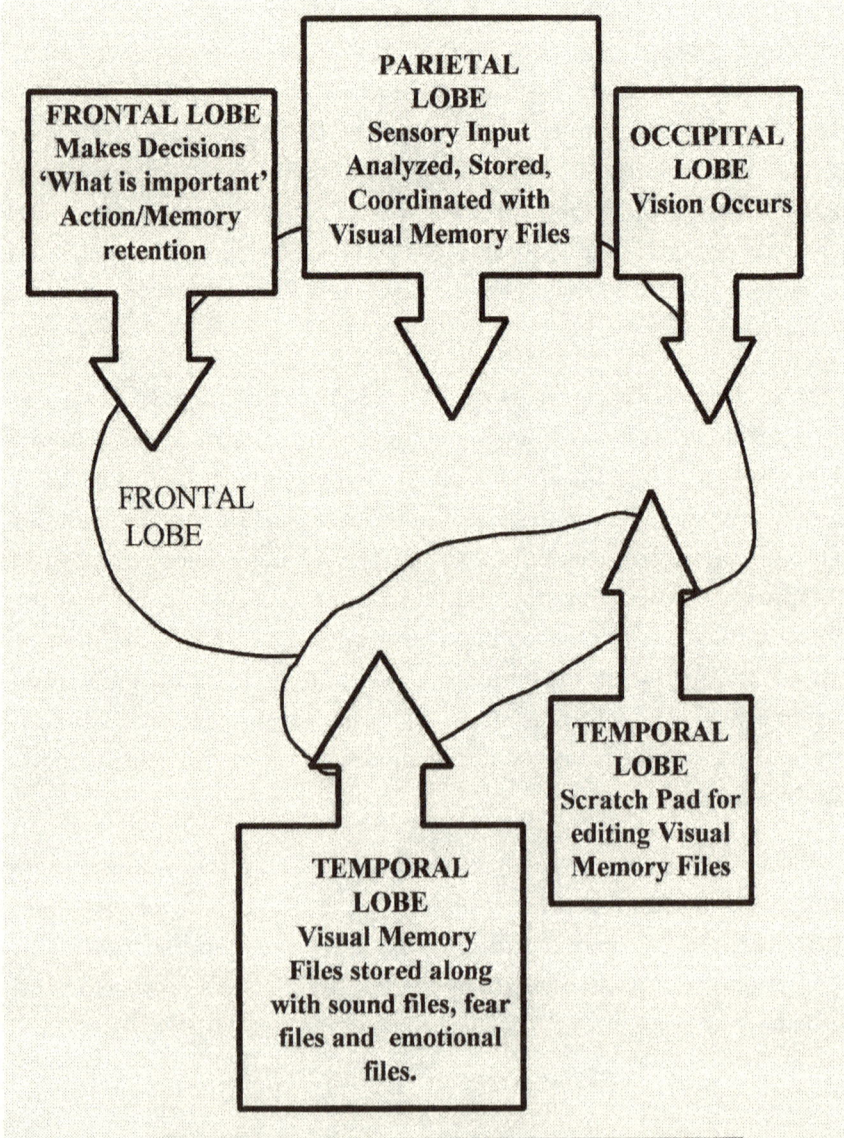

Figure 39. Human brain (left side view) areas where visual memory files are processed.

19

THE HUMAN NETWORK CAPACITY AIR, FOOD,...NETWORKING WITH OTHERS

Computers are networked through local area networks (LANs), they are networked to the Internet, they are networked as simple as when they send, or receive a fax. Networking computers together, to facilitate the ease of information transfer, is one of the more powerful assets computers have to offer.

Humans network today in many different manners. The communications industry is a testimony to human networking. Books, newspapers, televisions and movies allow downloading of information from one human to another. Phone systems allow us interactive two-way communication between one human to another. Cellular phones, pocket computers allow us wireless information transfer from one source to another. Recently, while on a medical business trip, strolling through the Atlanta airport, I realized how much we humans are already intimately networked together. Despite it being early on a Sunday morning, everywhere around the airport, other travelers had an earphone attached to their head and were actively talking to someone with their cell phone. I guess many of them were actively trying to use all of their allotted 3,500 free night & weekend minutes, advertised by some cellular companies.

The clear difference between computer networks and human networks is that computer networks do not send intact data streams from one computer to another. When a computer transmits information on a network, whether it be a local area network, or on the Internet, the information is broken down into data packets. The data packets are transmitted separately by the sending computer. The target computer receives the data packets and compiles the data packets back into the original data stream. For the most part, humans talk in data streams without breaking up the message. That is unless one is talking like: "Well,...ah...gees...ya,...but...ah...I...don't,...well, maybe...ah...ya ...sure George,...Come...ah...over...after school"; which then one might be considered conversing in packets of data versus in a single, coherent data stream.

The intention is not to poke fun at the concept of networking. In fact, like the desktop computer networks, humans networking their talents together, tend to be much more imaginative and productive, than an individual working on their own.

But, is there another, alternative means of networking? Is there a sixth sense, that in the new communication age, has been lost in the shuffle of stimulation and special effects? Do we have paranormal powers of telepathy? Being human, a biologic entity, comprised of trillions of cells, we exert an energy field. Some of us may be able to read the radiated energy more proficiently than others. Ever encounter someone, who instantly, from the first time you met them, you trusted them, and felt a bond of friendship? Or possibly the opposite, the hairs on your neck stood up and you sensed an instant distrust or danger?

The internal construction of our brains comprised of cisterns filled with cerebral spinal fluid (CSF), and the choroid plexus (strands of epithelium that generate CSF). The internal fluid-filled caverns, may offer the capacity to some humans of reading the thoughts or energy radiated by other humans. These fluid filled cavities inside our brains may act as a form of antenna that communicates directly to our brain, rather than through the five more tangible senses. There could indeed be validity to humans having a sixth sense, and because the detector is inside the brain, versus something we can touch and see as an external apparatus, only very few members of the population have learned to utilize and harness this sensory capacity.

If we knew more about how the brain functioned, and how we could best utilize its inherent features, the sixth sense might leap out of the pages of dark magic and witchcraft and into the arena of popular science.

IV. ESSENTIAL HUMAN COMPUTER FUNCTIONS

20

AWAKE, ALERT, AWARE, ALIVE, ASLEEP,...THE CIRCADIAN RHYTHM

The desktop computer is either on or off. If the desktop computer is 'off' it is about as inanimate as any object can get, except for a capacitor here and there that trickles life into a memory device or the battery charged internal clock. When the desktop computer is turned 'on', it acts as animated as the software that runs inside its plastic, silicon and metal body allows it to be.

By causal observance, it seems relatively easy to determine, what is alive versus what is not alive. Things that are alive, seem to move by their own means of locomotion. But then there are lots of things that are microscopic, that we never see. What microscopic structures are alive and which are not?

Amoeba are single-cell organisms that move, grow, and divide. Bacteria seemingly need only a slightly hospitable environment and they grow and divide, generally they do not move on their own power, but go with the flow of their environment. Viruses are simply genetic material wrapped in a protective shell. A virus must invade a host cell before it can grow and divide, and accomplishes this by inserting its genetic code into the genetic code of the host cell and taking control of the host cell's functions. So where does something cross the line of being alive, versus not being alive?

In scientific terms, structures that are considered alive, are constructed of biologic units called cells. Animal cells are comprised of a cell wall, cytoplasm, structures that float in the cytoplasm, and a very complex nucleus. Animals are made up of numerous cells. In higher-level animals the cellular units can be well-differentiated to perform specialized functions.

The human body is an intricate conglomeration of trillions of biologic cells. It is the coordinated effort of numerous, many highly specialized cells, that has lead to the success and survival of the human body in an otherwise very dangerous and demanding environment. Nerve cells, bone cells, skin cells, blood cells, and many other specialized cellular structures, add their unique functions together to comprise the human body.

ALIVE

The concept of exactly what being 'alive' is, is a concept that has eluded philosophers and biologists over the many ages humans have been able to ponder the matter.

One can re-ask the question, what is a computer? There is certainly a significant difference between a pile of computer components sitting in disarray on tabletop versus a computer organized and ready to accept and run an application program. There are many different types of equipment that could be categorized as having computer-like functions, though they may not have the full array of interactive functions that the latest desktop computer has to offer. A sprinkler system might possess some computer functions. The DVD player and newer television sets offer certain programmable functions. Many coffee pots offer a limited amount of programming options. Though these appliances are not in essence computers, they would be considered automated due to one or more computer chip(s) they are equipped with that facilitate programmable features.

Could one offer the definition that a computer is a device that is capable of running a computer program no matter how elementary the computer program is? The concept of automation refers to a self-operating machine or mechanism acting or operating in a manner essentially independent of external influence or control. By today's standard, 'automation' is often achieved by some form of computer operation.

If we stretch the imagination a little, and our definitions, the concept of being 'alive' may refer then to any structure that is capable of running a biologic program. Just imagine, being alive means the entity runs a genetic computer program...what an Earth shattering concept! How *intelligent* a life form is then, is determined by how sophisticated the biologic program is that the entity is programmed to run. The entity may carry a more sophisticated program in its genetic code than on what it runs or operates (thus the concept of evolution can occur and lead to sophisticated life such as humans); it is the program that the life-form runs that determines the entity's level of operating intelligence.

As humans venture off planet Earth and into the vast depths of space, our future means of defining 'life', in the infinite forms it may present itself to us, may be judged on the basis of whether or not the entity runs a computer program.

AROUSAL

Arousal requires the interplay of both the reticular formation and the cerebral hemispheres. The reticular components necessary for arousal reside in the midbrain and diencephalon. The midbrain may be viewed as a center that drives the function of the higher cortical structures. When midbrain function is lost, this produces a state where the cortex appears to become idle, waiting for instructions or commands.

AWARENESS

Awareness implies not only that an individual be alert, but that the individual is cognizant of self and surroundings. Communication between the cerebral cortex and the reticular system is required for an individual to be in a state of awareness.

ATTENTION

Attention is thought to be a state where by an individual is able to concentrate on specific aspects of one's surroundings. Attention depends upon a state of awareness and proper interplay of specific sensors and their associated brain structures. In order to attend to a particular stimulus, the pathways required for perception of the stimulus must be intact. For instance, the visual system must carry information from the retina in the eyes to the occipital cortex in the back of the brain for visual attention to occur.

ASLEEP

When our head hits the pillow at night and we fall asleep, most of us have a perception that the body is asleep. The reality is, that we are still breathing, our bowels are actively retrieving nutrients and minerals from the food we have eaten and dumping waste, our kidneys are actively producing urine, our muscles are attempting to relax in order to repair themselves and in men, at least, they have periodic erections—approximately five a night.

When we sleep, we tend to sleep in stages. These stages have been identified as stage one, stage two, stage three, and stage four. Stage four sleep has been termed 'REM' sleep, short for rapid eye movement sleep. In stage four sleep, the eyes flutter underneath the closed eyelids, thus the reason for the name. Stage four sleep is the sleep stage in which the muscles relax and, supposedly, the body reorganizes information it has learned in the past day. These four stages of sleep tend to cycle in forty-five minute intervals.

The point is, that though the conscious mind sleeps when we crawl into bed and place our head down on a pillow, there is a significant portion of our body that does not shut down, but remains quite active. The controller of these various bodily functions is the subconscious.

The human brain is comprised of at least three major functioning subunits which include the conscious, subconscious and memory. The conscious is the higher cortical function that operates in our wake state. The subconscious works continuously managing body functions. The memory stores vital information for later recall.

Our conscious brain works on all of the higher levels of thinking that are required to make us individuals, while the subconscious toils with all of the lower level thought processes that are required to insure the body functions properly. The subconscious interacts with the frontal lobes of the brain to work in a partnership. The subconscious takes care of most of the mundane functions of the body, releasing our upper cortical function to pursue *free thought*. That is, if our upper cortical function had to monitor every breath, or think about and effect a blink every time our eyes were dry, or monitor every muscle in our body, and properly reposition the body every few minutes to relieve pressure on the buttocks muscles while we sat and read this book, there would be no time to pursue independent, creative thought. Therefore, the thinking part of our brain is spilt into the 'conscious entity' and the 'subconscious' entity, or in other words the 'conscious' entity could be considered a patron staying at an exclusive vacation resort; while the 'subconscious' entity might be the manager of the vacation resort. The manager would oversee the details of the resort, making sure that the facilities of the resort functioned properly, so that the vacationer may enjoy himself or herself. The manager's concierge may expend time and effort locating information about the surroundings or solving problems for the hotel guests to make their stay as enjoyable as possible.

So, the subconscious might be thought of as similar to the central processing unit of the desktop computer. The subconscious could be termed the Human CPU. The subconscious also interacts with the emotional unit in the brain, the sexual drive unit in the brain, the logic unit in the brain and the peripheral devices located outside the brain.

The peripheral devices outside the brain include internal and external sensory devices, sound generation by the vocal cords, muscle function, and other bodily function such as devices to regulate body temperature, body blood sugar, blood pressure, body water balance, energy conversion and so forth. The external sensory devices include optics, audio perception, sense of smell, sense of touch and pain, the sense of taste.

The subconscious is also involved in problem solving. When the conscious brain cannot formulate an immediate solution to a problem, it may hand the duty over to the subconscious brain. Like a concierge at a prominent hotel, while the conscious brain is attending to the immediate needs, the subconscious brain toils with in-depth problems. Engineers do this all the time. They load their brains with parameters of problems, and let their subconscious work the problem. A solution to the problem may surface possibly minutes, hours, days or even years later, sometimes any time of day or night.

In simplistic terms, the human brain is comprised of a <u>conscious</u> that operates as our identity, the <u>subconscious</u> that constantly toils with making sure the body is functioning properly, and <u>memory units</u> that store information pertaining to our identity and our life experiences. The memory units allow us a place to transfer data files to so that we can manipulate the data and create new ideas. The memory unit is a storehouse information files passed down from generation to generation and as each day passes, our minds unlock new files from this storehouse that propel the human race forward.

A person's eyes suddenly flash open. The annoying ring of the alarm clock wails across the dark room announcing the stroke of six o'clock on the face of the clock. The person sluggishly rolls over and strikes the clock in a ritual that occurs every working day morning.

The body does have a twenty-four hour cycle referred to as the circadian rhythm, which is responsible for the production and timed release of hormones. Hormones are chemicals that help the body function properly. Hormones include the thyroid hormone, insulin, and sex hormones.

The Circadian Rhythm: Clear Representation of Human Programming

An hour before the alarm clocks rings to wake a person, the body begins preparing for the day to come. The circadian rhythm dictates the beginning production of the day's hormones. The heart starts to beat faster, the blood pressure increases. The glands begin to secrete hormones into the bloodstream. Though the alarm clock startles the upper cortical function of the brain, the body has already begun its preparation to wake up, much like a computer goes through a series of checks and organizational tasks when it is first turned on.

As the eyes flash open, the optic nerve, in conjunction with the lens located in both eyes, work together to attempt to put the visual acuity into focus. As we get older, this process takes longer and longer. In many of us, it's not until our hand locates the glasses on the nightstand next to the bed, and place the glasses on the bridge of our nose, that the eyesight comes into focus.

As you raise your head from the pillow and lift your chest, the sheets fall from your torso, the autonomic nervous system inside continuously recalibrates the blood pressure requirements to keep the brain perfused properly with the right amount of blood volume and pressure. Without this continuous monitoring and change in the blood pressure, you would become dizzy and pass out as you sat up.

But, you don't pass out, instead you yawn, and throw your arms up in the air to stretch your stiff muscles that have contracted during the night's sleep. You grasp the covers and toss them to the side. You kick your legs over the side of the bed. You shove your body's buttocks over the mattress to the edge of the bed and let your feet drop to the floor as you straighten your back and find yourself, once again, standing erect. Again, as you change position from sitting to standing, your autonomic nervous system struggles to regulate your blood pressure to perfuse your brain tissue properly, so you don't pass out. As soon as you start to move, the bladder in your pelvis and the intestines in your abdomen become stimulated. Pressure builds in the bladder and you search for the nearest toilet to allow the detrusor muscles squeeze the bladder of its urine volume. You struggle over to the shower and turn it on. You step into the shower and allow the hot water pelt your scalp and your skin. The warmth of the water and the striking of the skin by the water help to stimulate your sensors and wake your brain up. The warmth of the water is soothing, the striking of the water helps adjust the skin sensors.

The circadian rhythm (sleep-wake cycle) is a timed program that runs every day, in every human. The suprachiasmatic nucleus (SCN) of the hypothalamus functions at the biologic clock for the brain. The SCN has a natural cycle of slightly greater than 24 hours, but is reset to the 24 hour rotation cycle of the planet by influences caused by sunlight detected by the photoreceptors of the retina in the eyes and transmitted by way of the retinohypothalamic pathway to the SCN. In instances where an individual is isolated from exposure to sunlight, they will adopt a daily rhythm of 24.5 to 25 hours duration.

The circadian rhythm is responsible for regulating resources for locomotion, food and water intake, sexual behavior, hormone levels, core body temperature and creating a rest phase. Hormones are released during the 24 hour day cycle at pre-programmed times. The hormone cortisol that helps stimulate body function, is released in its highest concentrations between 4:00 to 8:00 in the morning. Thyroid stimulating hormone levels, which stimulates the thyroid to secrete thyroid hormone, a master regulatory hormone, increase at about 11:00 at night, just before sleep. Many other hormones are released at predetermined times during the day to create an efficient use of the body's resources.

To recognize how powerful the programming of the circadian rhythm is, one simply has to take a transmeridian plane trip that crosses three or more time zones. When the body's internal rhythm becomes out of synch with the external environment, an individual tends to experience what is termed jet lag. Recovery or complete readjustment to the new environment may take up to seven days.

◆   ◆   ◆

The circadian rhythm is one of the recognized features of the human body that clearly demonstrates the human brain is a programmed biologic structure.

◆   ◆   ◆

As I get older, I realize that I am no longer waking up in the morning and instantly ready to go, as I did in my teens, twenties and thirties. Within the first half hour every morning, it is taking time now for the brain to wake up, the eyes to adjust themselves and for the stiffness in the muscles and joints to work itself out. I often feel like a computer when you first turn *it* on. My desktop takes a good five minutes after I turn it on to get itself together before it's useful.

My younger days were quite different. I remember one night in the hospital. I was designated the senior resident, which meant that in addition to overseeing a team of interns and their patients, I was required to respond to and lead emergency situations that might occur anywhere in the hospital. You slept when you could at night, but often you didn't get much sleep. It was routine to work all day, either in the hospital or the clinic, then work all evening and night, and then without going home, or having any designated break, work all the next day in the hospital or the clinic depending upon the rotation assignment.

One particular night, I remember having the chance to catch a few minutes of sleep. It was one of the few nights I had a chance to see the inside of the call room. The call room, as it was referred to, was simply a patient room that was designated to be used by the on-call residents. When you were the senior resident, you slept in your scrubs and with your shoes on. You placed the code beeper somewhere near your head.

Somewhere around three in the morning, I remember the code beeper going off with a gut-wrenching wail. A mechanical voice blurted over the pager's small but feisty speaker, calling out, "Code Blue F4- Room 36." The voice always repeated itself one time.

I remember blasting my head up off the pillow like a rocket shot into outer space. Since I was already dressed, with my shoes on, I remember leaping out of bed. My head was fuzzy, I didn't quite know who I was or even where I was. Responding to the code beeper was more of a nervous reflex. I recall finding myself opening the door to the room and bolting down the dimly lit hospital hallway. I remember thinking to myself that my legs were running, that I was headed in the direction of the nearest stairwell (we never relied on the elevator, we always used the stairs to respond to a code) and yet, I distinctly remember that my conscious mind was not awake enough to know where I was, much less who I was or what I was doing. My legs seemed to know where I was headed, but my conscious self didn't. Because a code blue represented a life and death situation for one of the patient's in the hospital, following the code beepers being activated, overhead the loud speaker system would blare out the location of the code. It was finally after the second announcement over the load speaker that I recognized who I was and the location in the hospital complex that the code team had been called to.

Following this experience, I realized that (1) you didn't necessarily have to be awake to be running down a hallway and (2) I was thirty-one years old and I no longer was capable being woken up and immediately ready for action…It took a few moments to get my human computer system up and running and <u>fully</u> functional. Thank God my subconscious was well in control those first few steps, as I clambered down the hallway, before my consciousness shook itself from the clutches of sleep that clouded my head.

21

THE AWESOME PROCESSING POWER OF THE HUMAN BRAIN

At the core of conventional computers resides machine language. Machine language is based on the mathematical language of ones and zeros. As far as a conventional computer is concerned, everything is either 'on' or it is 'off' (in a measurable sense at a potential of 5 volts for 'on' and at zero or 0.3 volts for 'off'). In addition, up until recently, computers were based on 16-bit processors. That is, at any given time only sixteen ones or zeros or some combination of ones and zeros could be transferred at a time around the computer through the information buses. That meant that instructions and data were limited to 16 bits at a time. Newer computers are constructed with 64 bit processors and information buses that can transfer as many as 64 bits of information or, in other words, sixty-four ones or zeros or some combination of ones and zeros in/or out of the processor via the information buses at any given time. Though the computer processing power may improve, a limiting factor to conventional computers is the software to run on the computer. Much of the software to date is designed to run on 16 bit or 32 bit computers, and therefore, may not take full advantage of the increased processing power a 64 bit computer might offer. Future computers may run with 128 bit, 256 bit or larger central processors and data buses.

Intel's Pentium III processor has 9.5 million transistors. Intel's Pentium 4 processor has 42 million transistors with 15.5 million dedicated to a high-speed internal memory cache.

The human brain is comprised of 20 billion interneurons. Interneurons are nerve cells dedicated to processing information inside the scope of the brain. There also exist peripheral nerves that send signals from the numerous exterior sensors to various locations in the brain, and motor neurons that relay commands from the brain to various muscle tissue and internal organs, in order to regulate the actions of the body.

The extensive processing power in the small confines of the human brain is, in part, due to the fact that unlike conventional computers, the human brain is not limited to the binary number system that current desktop computers are limited to.

The processing power, the speed of the brain and the memory capacity outperforms any computer we have available today or in the foreseeable future. The secret behind the processing power of the human brain lies in the nerve cells that traverse the brain. Where two nerves meet is termed a *synaptic junction* or alternatively a *nerve synapsis*. Actually, more than two nerves may merge together at a single nerve synaptic junction.

The Chemical Versatility of the Nerve Synapse

Nerve signals are conducted along the lengthy body of the nerve by an electrical impulse. At the terminal end of the nerve, the signal traverses across the nerve synapsis by means of a chemical reaction. When a chemical reaction occurs inside the nerve synapsis, the chemical reaction can up-regulate or down-regulate one or more of the outgoing nerves or efferent nerve leaving the synaptic junction. The outgoing nerve or nerves carry an electrical impulse through the nerve body if up-regulated. If the chemical reaction causes the nerve to be down-regulated, the outgoing nerves may not respond to other signals entering the nerve synapsis until the chemical reaction present in the nerve synapsis ceases or is countered by further chemical reactions (see Figure 40).

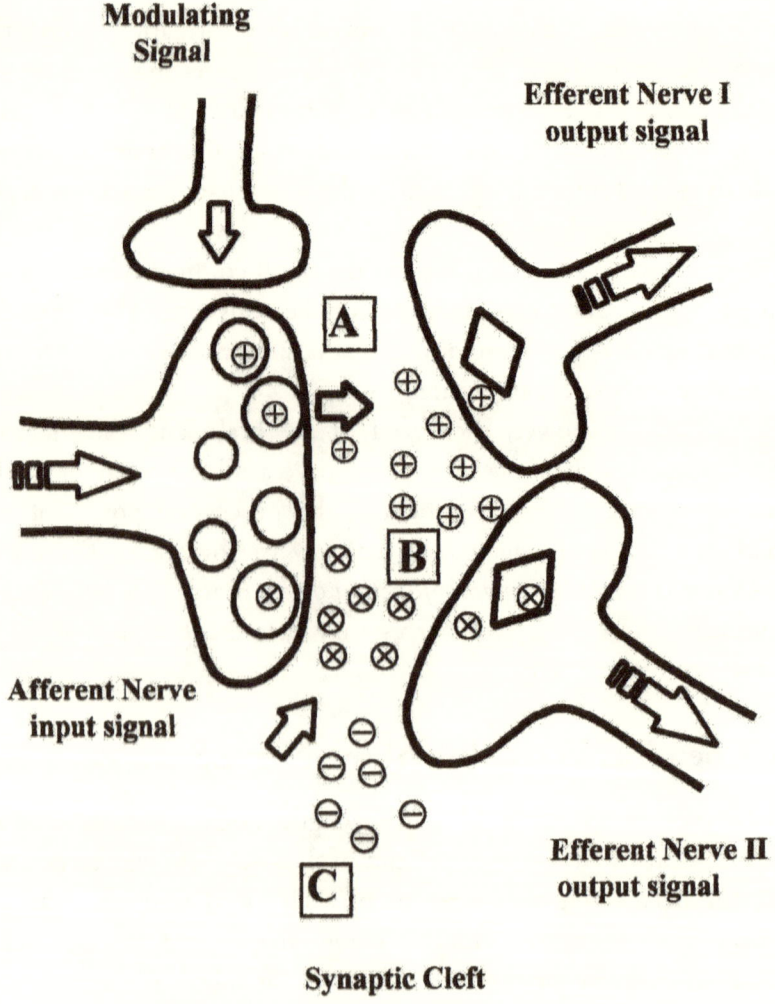

Figure 40. A nerve synapsis demonstrating multiple input signals and output signals

A chemical reaction that occurs in the terminal end of a nerve usually involves the rupture of a vesicle that holds a neurotransmitter molecule. When stimulated properly, an afferent nerve ending will cause a vesicle to release neurotransmitter molecules into the synaptic space. These molecules will traverse the nerve synapsis and stimulate the other nerve endings comprising the neuro junction.

Regulation of the transmitted signal is effected by the amount of neurotransmitter released into the neuro junction, the presence or absence of a neutralizing agent, the length of time the neurotransmitter is present before neutralized, the level of receptibility of the target neuron (type of receptor), presence of competing neurotransmitters in the neurojunction that may block the effects of the a neurotransmitter by binding to the target receptor before a specific neurotransmitter is able to bind to its target site.

Figure 40, demonstrates a generic nerve synapsis. There may be one or more afferent nerves endings that terminate at the junction. There may be one or more efferent nerve endings that originate at the nerve junction. There may be one or more nerve projections, referred to as a dendrite, that connects to a nerve to up-regulate or down-regulate the afferent nerve signal. One or more transmitter chemicals (see A and B) might be released into the synaptic junction in an attempt to stimulate the efferent nerve(s). One or more chemicals may flood into the nerve synapse (see C) to block or down-regulate the afferent nerve's transmitter chemical(s), or stimulate the afferent nerve to continue to release transmitter chemical, or artificially stimulate the efferent nerve ending to produce efferent nerve signal.

The well-known disorder Parkinson's disease, is a condition where portions of the brain does not generate enough Dopamine. Dopamine is a neurotransmitter found in a number of areas including the red nucleus of the brain. When the brain is lacking in adequate amounts of Dopamine, body movements become rigid, facial expressions blank, and repetitive motions of the fingers referred to as 'pill rolling' may become apparent. Treatment of Parkinson's has included providing patients with L-Dopamine, in an attempt to supplement the Dopamine the brain should be making.

At the nerve synapsis, multiple chemical reactions may occur at one time. Each electrical impulse traversing down an incoming nerve, may result in a different chemical reaction. These different chemical reactions may result in the release of different neurotransmitters. These neurotransmitters travel across the synapsis junction and stimulate the outgoing efferent nerve(s) in different ways. Since it is possible for a number of different neurotransmitters to be released from the incoming nerve and absorbed by the outgoing nerves, a single nerve synapsis can function in different capacities, simultaneously, resulting in different signals traversing a single nerve junction. In addition to neurotransmitters acting separately from each other, neurotransmitters released into the neurojunction may either compliment or antagonize each other.

Therefore, in the human brain, instead of a nerve junction acting as a 'one' or a 'zero', or acting as simply 'on' or 'off', a human nerve synapsis may communicate a number of signals, all at once, to the outgoing nerves tied to a nerve synapsis, by utilizing a number of different neurotransmitters.

Due to direct-to-consumer advertisement, in recent years, the general public is becoming aware of medications to treat depression and other mental illnesses based on either adding or inhibiting neurotransmitters. Older antidepressants worked as MAO inhibitors. MAO inhibitors block monoamine oxygenase in the brain, which is thought to be beneficial. Most recently, serotonin re-uptake inhibitors, are being advertised. Serotonin is a neurotransmitter, and the re-uptake inhibitor medications simply cause the serotonin to remain in the neurojunction longer. The awesome power of the human brain is, in part, due to the versatility by which the nerve synapsis work.

The Base Four Power of DNA

Another simple example of how advanced the human computer is in relation to our conventional computers is seen in investigating the chromosomes that exist in our cells. Chromosomes are long strands of DNA (deoxyribonucleic acid), see Figure 41. DNA is constructed of two strands of nucleic acids coiled into a double helix. The nucleic acids comprising the strands represent data storage. There are three billion base pairs that make-up the human genome (genetic data library).

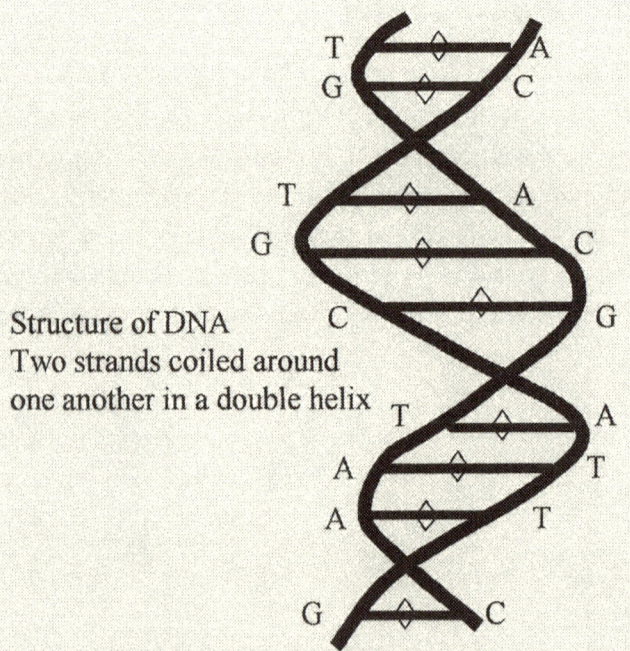

Structure of DNA
Two strands coiled around
one another in a double helix

Each segment is made of one purine
connected to a pyrimidine by hydrogen
bonds:

4 Nucleotides

Purines   **Pyrimidines**
A = Adenine   T = Thymine
G = Guanine   C = Cystine
Adenine is always paired with Thymine and
Guanine is always paired with Cystine.

Figure 41. Human DNA structure made up of
a nucleotide, either Adenine, Thymine,
Guanine, or Cystine, and this produces a base
four genetic code for storing information.

In the human, there are twenty-six pairs of chromosomes. One pair of the twenty-six pairs of chromosomes are the sex chromosomes. The sex chromosomes are comprised of either two 'x' chromosomes which result in an individual with female organs and female physical characteristics, or an 'x' chromosome paired with a 'y' chromosome, which results in an individual with male organs and male physical characteristics.

These chromosomes are stored in the nucleus or intelligence center of human cells and provide the memory information required, not only with regards to how to design and construct a human body, but how the human cells are to run to maintain the human body once it has been constructed.

Again, where conventional computers memory chips are made of transistors that when *energized* represent a 'one' and when *not energized* represent a 'zero', the chromosomes are made up of four different base pairs. Therefore, where conventional computers use a binary system, the chromosomes use a quaternary system. In terms of memory capacity, the conventional computer is limited to being in a state of either on or off. Utilizing a quaternary memory capacity, there are four choices to be used in combination.

Chromosomes are made up of long strands constructed of many paired genes. There exist four different types of nucleotides that make up these genes. Two of the nucleotides are purines: Adenine and Guanine. Two of the nucleotides are pyrimidines: Thymine and Cystine. In the DNA structure, the nucleotides act as rungs of a ladder. For each rung of the ladder two nucleotides are paired up with hydrogen bonds between them. Adenine and Thymine are always paired together. Guanine and Cystine are always paired together. This pairing produces rungs that are the same width, and therefore the two backbone strands of DNA are always the same distance apart.

Since one of four different nucleic acids exist at any given point along the chromosomal DNA strand, there exists one of four possibilities. Arranging the nucleic acids into groups produces packets of information, which are referred to as genes. Grouping just three nucleic acids together offers 64 different combinations, where as in the desktop computer's binary system, three transistors would only allow 2^3 or eight different combinations.

The human brain is a multi-leveled organ, operating in many complex capacities, the secrets of its inner workings continue to mystify medical science.

Still, the power of the human brain is, in the fact, that the number of neurons is approximately 2,000 times greater than the Pentium IIIs number of transistors, and four hundred times greater than the number of transistors in a Pentium IV. The brain uses as many as sixty different chemicals in order to transmit signals between neurons. The desktop computer uses only one medium, the presence or absent of an electric charge in a transistor, and the flow of electrons to transmit data. The desktop microprocessor, in some cases, uses the same bus to bring information into the CPU and to move information out of the chip. The brain has dedicated pathways for information arriving into the brain and for sending command information out of the brain to muscle and organ tissues. The brain can be transmitting and executing commands while it is in the process of accepting new data. There are many features of the human brain's processing power that offer us clues as to how to improve desktop computer technology in exciting new ways.

Stereoarchitecture

The blueprint information stored in the chromosomal DNA may lead to a specific way the neurons are created in the brain, not only in a gross sense as what has already been explored in the boundaries of this book, but also in how decision trees are physically created amongst the neuro-network.

A computer chip is created by a team of designers. The computer chip is manufactured in an automated manner. The parameters of the chip's design are fed into manufacturing computers that construct the chip on an assembly line. A specific design leads to a computer chip with thousands or millions of transistors arranged in specific spatial relations inside the final chip, which correspond to decision trees or data storage.

In the human brain, a similar concept may apply. The chromosomal DNA may lead to the specific spatial design of decision trees and data storage in the human brain. Generation after generation pass along similar capacities to think, create, and problem solve due to a similar decision tree structure that occupies the human brain. How the components of the brain are hardwired may also confer different means of processing. In addition, generation after generation of humans might carry pre-existing data files in their brains, filled with facts regarding a whole variety of subject matters, that come to general knowledge as either a need arises, or timing and environmental factors trigger the opening of a data file stored in an untapped section of the brain.

22

THE MIRACLE OF SIGHT

Color digital cameras and video cameras are constructed with strips of photodiodes located behind the camera's lens. Covering the strips are blue, green and red filters. When exposed to light, the photodiodes convert the amount of light they receive to a corresponding voltage. This voltage is transferred to analog-to-digital converter (ADC) chips. The ADC turns streams of analog data into digital data color values for each photodiode. This data is then either transferred to a memory device or a video screen for viewing.

Human sight is a complex biologic function, and unlocking the secrets of sight might help us understand how the human computer transfers data.

Light strikes the front of the eye. Light passes through the lens in the eye and enters the inner chamber of a round globe. Inside the globe of the eye, the eyeball is filled with a clear fluid. After passing through the fluid inside the eye globe, light strikes the back of the eye, activating the retina. The retina is rich in optic sensors which capture light energy. Light focused by the lens stimulates the optic sensors, first order nerves. The optic sensors turn the light energy into chemical energy.

The chemical energy generated by the photosensitive nerve cells in the retina is transferred to the optic nerve (see Figure 42). The signal is transmitted via the optic nerve posteriorly to the back of the brain. Midway, some of the fibers of the optic nerve cross to the opposite side of the brain. This crossing of fibers generates a means whereby, if a visual object is to a person's left, then the right side of both retina in both eyes will detect the image and transfer the information from both eyes to the right side of the brain. If an object is to the person's right side, then the left side of the retina of both eyes will detect the object and transfer the information to the left side of the brain for interpretation. From the optic chiasm the optic nerve travels back to the geniculate nucleus on either side. The geniculate nucleus nerve signals are transferred along a span of fibers, termed the optic radiation, to the visual neurocortex in the occipital lobes, which reside in the back of the brain, and interpret the light signals.

The retina is comprised of two major types of optic sensors. Rods are used to detect daytime light. Cones are used in nighttime vision to detect light images. The retina is comprised of 120 million rods and 7 to 8 million cones. Several layers of nerve cells are located between the rods and cones, and the optic nerve. These nerve cells assist in the processing of light images.

Vision is not a continuous function. Since detection, processing and transmission of neurosignals are chemical functions, there exists a delay in signal transmission. Light images are captured by the retina, a chemical signal is produced, then the chemical signal decays as photosensitive molecules reset themselves. Light images are sent down the optic nerve to the occipital lobe of the brain in discreet packets of information.

The occipital lobe receives packets of data. There exists a persistence of the image due to the optic sensors requiring time to reset themselves. Since a persistence of an image exists, light sources that are noncontinuous may appear continuous to the human brain if the light source flickers faster than the brain's integration time. A theater's movie projector shutters operate faster than the brain can interpret the light signal. If a movie film runs at 24 frames a second, and the movie projector uses a triple shutter, then the flash rate is 72 times a second. The human brain interprets the flashing movie frames as a continuous image. Computer screens flash at least 75 times a second, therefore, also appearing as a continuous signal to the slower human visual processing capacity.

In addition to image signals being transmitted from the eyes along the optic nerve to the occipital lobes at the rear of the brain, a portion of the light data information is transmitted to the suprachiasmatic nucleus (SCN) along nerve fibers of the retinohypothalamic tract. The SCN is the biologic clock. The SCN resides in the hypothalamus, positioned just behind the optic chiasm (where the optic nerves cross fibers). The body's circadian rhythm is dependent upon programmed impulses generated by the SCN. The SCN has a pre-programmed rhythm of slightly over 24 hours in duration. Light impulses transmitted to the SCN, when the eyes are exposed to bright sunlight, help to reset the SCN impulses to the 24 hour rotational cycle of the planet, to keep the body's internal rhythm synchronized to the external environment.

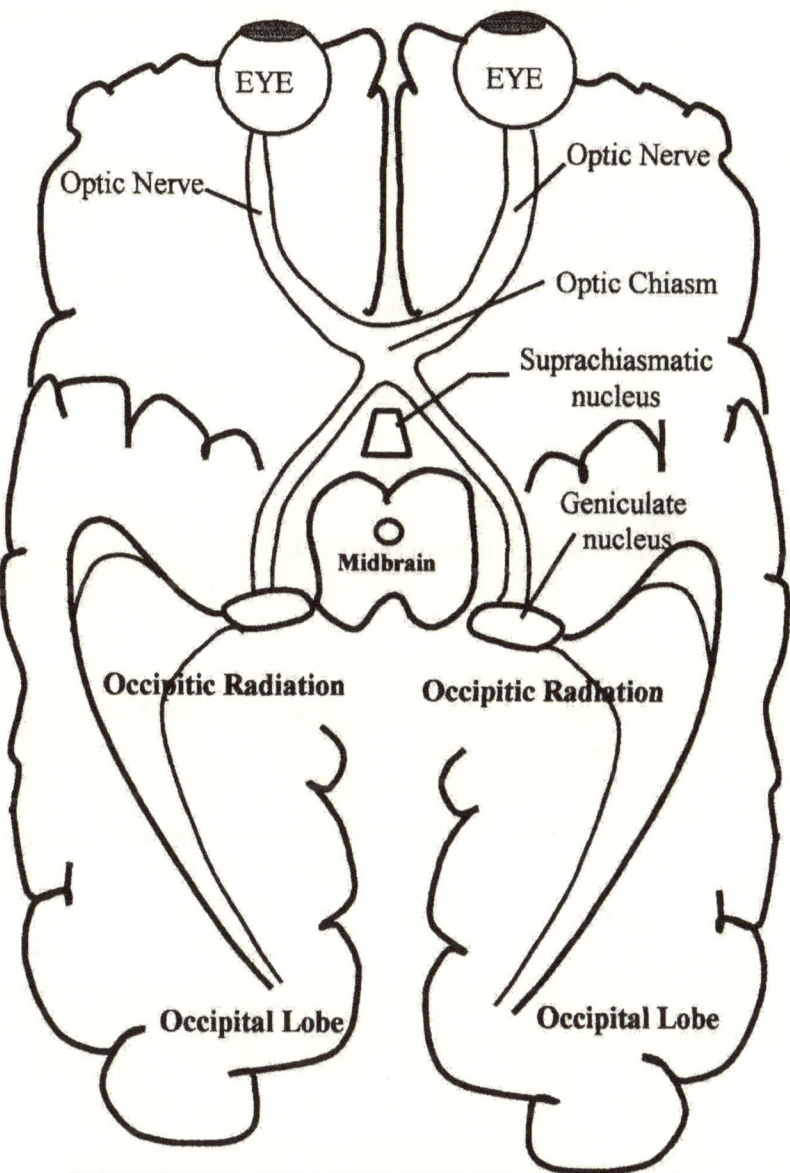

Figure 42. View of the Optic system from on top

Alternative manners of sight

There exists two alternate means of vision. These two alternative means of vision may be referred to as *tactile* and *recalled* vision. These two alternate forms of vision are partially related to data input by the eyes, but are not dependent upon real-time line-of-sight vision (what the eyes are directly looking at). The two alternative forms of vision can take one or more images previously seen by the eyes, and recall the image(s) and utilize or change the image as required by the individual, possibly adjusting the image using sensory data input by the tactile, temperature, vibration and/or proprioception sensors.

As a physician, certain parts of a patient exam, I often will look away from the person or close my eyes. I have found, that in the physical exam of a breast, abdomen or an extremity, the tactile inputs sensed by the fingers and palm of the hand create imagery in the brain that is sometimes clearer than what my eyes can see. The senses of touch, temperature, vibration and proprioception combine to create in the brain a three dimensional spatial image of the part of the body being examined. Often the eyes can only see the two-dimensional image, where the hands can create a more detailed three-dimensional image.

As a plumber, and working as a car mechanic, often times completing a task depended upon tactile vision. Many projects required getting the hand or simply one or two fingers in a tight spot, and accomplishing some task, whether it be ratcheting a bolt, gluing a pipe, screwing on a nut, possibly hammering a nail. Often plumbing and engine repair tasks were out of the direct line of sight. The only way to accomplish the task was letting the fingers sense the area surrounding the task, and develop imagery in the brain that corresponded to what the fingers felt. Then, the task could be accomplished by relying solely on the information the fingers conveyed to the brain, without ever seeing the physical form of the task with the eyes.

The part of the brain responsible for coordinating the data input from the tactile, temperature, vibration and proprioception sensors on the skin is the posterior (rear) portion of the parietal lobes located behind the central sulcus (the depression that exists between the front and back halves of the parietal lobe). The rear portion of the parietal lobe borders the occipital lobe (located in the very back of the brain). The parietal lobe-occipital lobe junction is the area likely where the imagery created by tactile stimulation is processed, and projected so that the conscious brain can 'see' and interpret the imagery.

An additional, alternate means of 'seeing', without directly utilizing the real-time imagery seen by the eyes, is referred to as *recalled vision*. We use recalled vision all the time. Whenever one reads written text that describes a physical object or place, the mind creates an image that the conscious brain can 'see' and interpret. The image created is comprised of details derived from image memory files already accumulated in the brain given the experiences the individual has had.

If an author were to write, a red brick house, decorated with pristine white shutters, topped with a slate gray roof, resting on a flat, lush green lawn with an ivory white picket fence stretching across the perimeter, cut by a cobble stone walkway gently snaking across the front lawn up to the front door, the reader would 'see' this image created in their head without actually 'seeing' the house. Such recalled imagery can be changed. The above description of a house could be modified to read, a red brick house, with violet colored shutters, topped by a black roof, resting on a flat, sunburned, crispy brown lawn with a white picket fence desperately in need of re-painting, stretching across the front lawn with a cobble stone walkway snaking up to a plum purple front door. The recalled memory can be modified to fulfill the creative needs of an individual, whether it is a function of their career, or a hobby that entertains the person, or an upgrade or improvement to the surroundings where one lives.

The detail of the recalled memory is dependent upon the visual experience of the person. If one were to write *the Jollyweg that lived in the sloop invested wriggles of Vega*, one would not be able to conjure up an image for this description because no one from this planet has visited the star Vega. On the other hand, many readers would be able to 'see' in their minds the image of a baby elephant, if it were announced, let's say on he radio, that a new baby elephant had been born at the local zoo.

The recalled visual memory is vital to our ability to communicate with others. Being able to listen or read what another individual has said or written, and then able to 'see' in our own minds what it is the individual is communicating, without actually seeing with our own eyes what they saw, allows us to think in the abstract. Being able to think in the abstract, and recall previously saved image files and to be able to manipulate these image files leads to much of our individual creativity, design effort and artwork.

The recalled memory could be considered the *scratch pad* memory, because, like a scratch pad, in this area of the brain, one can recall a memory file, revise the imagery of the file, and then chose to store the memory file as a different memory file, or erase the memory file (forget it). The projection area (where the imagery is seen and interpreted by the brain) is likely located in the posterior (rear) of the temporal lobe at the border of where the temporal lobe and the occipital lobe meet. The temporal lobe facilitates storage of long-term memory. When the conscious brain in the frontal cortex, wishes to recall a visual memory, it requests the temporal lobe to access the memory file, then project the file to the temporal lobe-occipital lobe border so the consciousness can 'see' (interpret) the file.

One, two or even all three of the visual imagery projections can be at work simultaneously. One can be looking at an object, projecting the line-of-sight image to the occipital lobes by means of light impulses from the eyes transmitted to the occipital lobes by means of the occipital nerve fibers, while one is either working with their hands and/or creating an entirely different image of an object in their scratch pad memory. When more attention is given to the activity occurring in the scratch pad memory then what one is looking at with their eyes, the individual might be labeled as daydreaming. An example of this form of brain multi-imagery-tasking might be the instance where a teen is lying on his back with his left arm stretched up under his car. While the individual's left hand is attempting to screw a bolt into a hole that cannot be seen, to secure the car's oil pan to bottom of the engine, the individual is also peering at his friend sitting next to him, and as the two talk about the high school dance they intend to attend that night, the individual lying on his back conjures up the image of a girl he is interested in, that he hopes will attend the dance. It is obvious, that at an early age, we learn to multi-task our imagery potential.

Understanding better how the brain takes visual signals and encodes them so the data can be transferred to the back of the brain and to other structures such as the SCN, and how such signals are interpreted is a key to unlocking the programming language of the brain. There is much to learn. The eyes are the window to understanding and communicating with the human conscious and subconscious computer, and all of the processing features of the brain.

23

HUMAN STEREO SOUND SYSTEM

As technology advances, so does the notion of humans interfacing with machines during their daily lives.

Science fiction would lead us to believe that jobs such as a library or museum attendant, or an office receptionist, might be replaced by holographic images or mechanical robots that would be capable of directing humans to destinations, answer questions, and providing background information regarding exhibits. Sophisticated computer animation may be able to program in facial expressions as a part of the response to human requests.

From a computer science perspective, the steps required to accomplish creating an animated robot or holographic image requires a means of gathering data, performing voice recognition, deciphering the gathered data input per a series of decision trees, then responding to the human's verbal requests with appropriate speech including grammar, wording, and punctuation.

To design an interactive robot or holographic image would require a microphone to gather acoustic data. The acoustic data would need to be digitized and background noise filtered out. Multiple requests posed to the robot simultaneously, by different humans would have to be isolated and the decision to respond to one of the requests at a time would need to be made. The selected request would then be fed to the voice recognition software. Once the wording was deciphered, the speech recognition program would feed the data to the language decision trees to discern the meaning of the human's request. Once the meaning was deciphered, an appropriate response would need to be generated. Once the response was created, the language decision trees would be utilized in the reverse, to create a string of words in order to create a verbal response. The response would be transferred to a speaker, and a computer-generated voice would be made audible to the human interfacing with the artificial intelligence system. In addition, expressions that correlated with the meaning of the wording would need to be generated.

The human brain performs similar functions in order to accomplish daily speech. Sound is captured by the ears. The sound is channeled by the inner chamber and focused in the direction of the eardrum, the tympanic membrane located at the end of the canal. The vibrations that resonate on the surface of the eardrum are transmitted to the three ossicle bones that sit behind the eardrum in the middle ear. These three very small bones, the malleus connects to the incas, that connects to the stapes. The energy is transmitted through the three ossicle bones to a membrane. The sound waves from the membrane are transferred to a fluid filled structure known as the cochlea (see Figure 43).

The cochlea is a spiral shaped structure that deciphers the volume and frequency of sound and transfers the data into nerve impulses. The sound data is transmitted from the first order neuron, the cochlea nucleus, via the eighth cranial nerve to the second order nerve nuclei located in the brainstem. The second order nuclei, known as the eighth nerve nuclei, sends audio information up to the cerebral cortex via the lateral geniculate nucleus. At two levels, nerves cross, so as to supply both sides of the brain with the same acoustic data.

It is curious that the brain is constructed with the acoustic data being shared by both sides of the brain. It is certainly advantageous that the brain shares acoustic information with both sides of the brain, because in the event the acoustic nerve on one side of the head fails, the individual may still feel they are hearing from both the affected side and the nonaffected side and, therefore, may not appear in social situations, to be actively compensating for the deficit.

From the lateral geniculate nucleus, the acoustic data is transmitted to Wernicke's area, located in the posterior temporal region of the brain. In Wernicke's area, acoustic input data is deciphered and speech recognition occurs. Nerve impulses are sent to the frontal lobes, by way of a nerve bundle (data bus) referred to as the arcuate fasciculus. If appropriate, a verbal response is generated in the Broca area in the inferior frontal lobe.

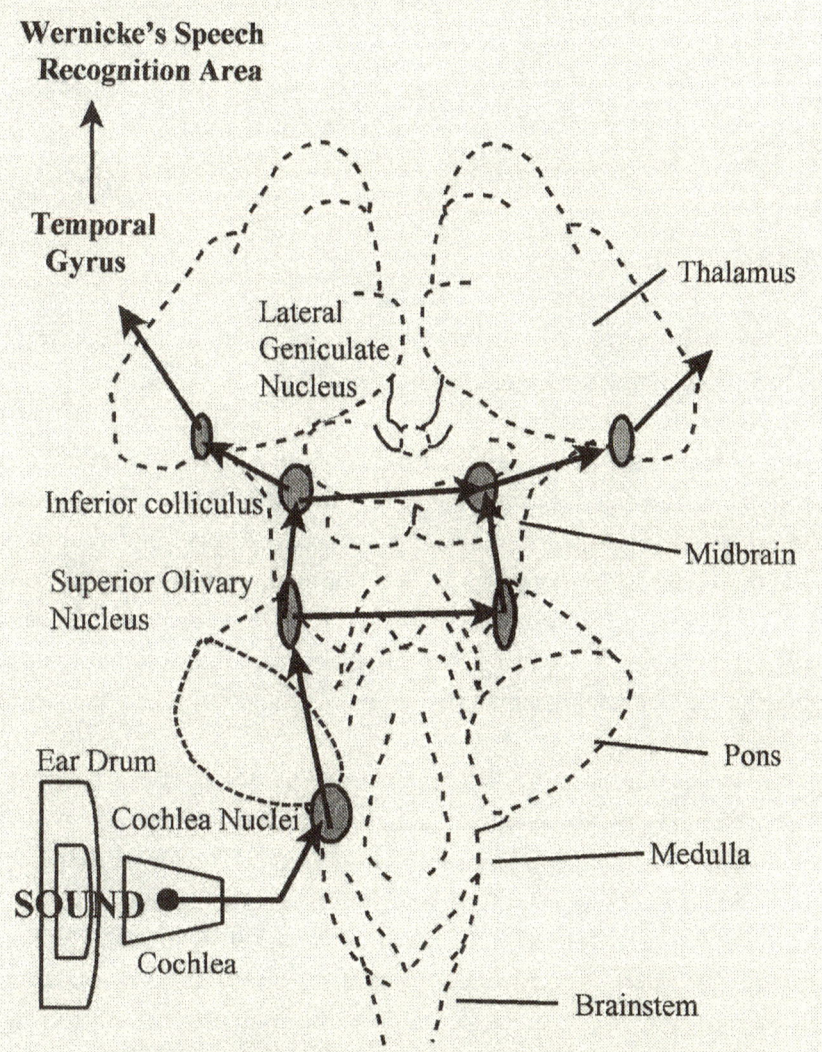

Figure 43. How the brain hears: Sound is transferred to electrical impulse at the cochlea, which is sent to the cochlea nerve, then to the superior olivary nucleus, then inferior colliculus, then to the lateral geniculate nucleus, then to the temporal gyrus (front view).

24

TELL IT LIKE IT IS

Through a microphone, sound enters a desktop or tower computer. The analog signal generated by the sound is sent to an analog-to-digital converter (ADC) often located as circuitry on the sound card (sometimes located on the motherboard). The now digitized audio signal is then stored in the computer's memory as a sound file. The sound may initially be stored in RAM (random access memory) and then sent to the hard drive or rewritable CD or a floppy disc for long term storage.

To play an audio file, the memory file is sent to the sound card. A digital-to-analog converter (DAC) converts the digital signal to an analog signal. The analog signal is then fed to the computer's speakers and sound is generated.

The way humans speak to each other involves several parts of the brain working in concert. Hemispheric dominance is not designated by which hand is your dominant or strongest hand, but by the side of the brain in which comprehension and production of speech occurs.

As previously described, humans detect sound with their ears. The cochlea transfers sound to nerve impulses. The cochlea is the human's analog-to-digital converter. The nerve impulse is sent along the eighth cranial nerve to the cochlear nerve nuclei in the medulla. The sound signal is sent to the superior olivary nucleus, then to the inferior colliculus, then to the lateral geniculate nucleus, then via a nerve pathway termed the auditory radiation, the sound signal is transferred to the interior of the temporal lobe of the brain. Hearing occurs in the central portion of the temporal lobe. Long-term memory occurs in the posterior (back) portion of the temporal lobe. Storage of auditory information occurs in the region where the temporal lobe and the occipital lobe merge in the dominant hemisphere. Wernicke's speech area is located in the area where the temporal lobe attaches to the main body of the brain at the lower section of the parietal lobe. In Wernicke's speech area, the human brain recognizes and understands language sounds (see Figure 44).

To generate speech, the human uses the Broca speech area located in the forward section of the brain, in the lower portion of the frontal lobe of the brain. Language (words and grammar) is generated in the Broca area, and the information is transferred to the lower mid parietal lobe. At the lower end of the central sulcus, the movements of the mouth and tongue are designed. This acts as the digital-to-analog converter of the human speech center. This lower part of the parietal lobe generates command signals that dictate muscle motor function that are then transmitted down through the diencephalon, to the brainstem. In the brainstem, nerve fibers of the fifth, seventh, and twelfth cranial nerves work in concert to create speech, orchestrating movements of the mouth and tongue and facial muscles.

From the seventh nerve nucleus, nerve fibers transmit the motor signals along the seventh nerve to the mouth, tongue and vocal cords. The combined function of expulsion of air from the lungs, motion of the vocal cords located at the top part of the trachea, and motion of the tongue and mouth creates speech. In addition, movement of certain facial nerves creates expressions on the face of the human, which may help the person to whom the speech is directed, understand the meaning and/or emphasis of the spoken language response.

A data bus (the arcuate fasciculus) projects from Wernicke's speech area in the rear of the brain, to a data bus in the front of the brain (the superior longitudinal fasciculus) which terminates in the Broca area located in the front of the brain. 'Language' to the human brain, is much like an application program or due to the complexity and necessity, what an 'operating system' is to a desktop computer. Language facilitates audio communication between humans, efficiently and effectively. Some humans are capable of learning more than one language, and therefore, work with several operating systems. They are able to take the same concept file and attach more than one word or expression to the concept. Many people perform a similar function, when we memorize different terms for the same *word meaning*. Additionally, individuals that learn more than one language are able to string together word files using different grammatical rules. That is, in the Broca area where speech is formed, they will arrange the word files in a different order if they are speaking English, than if they are speaking another language such as German or Spanish, in order to conform to the rules of how speech is generated, so that they can be understood by other humans speaking the same language.

Aphasia is considered the partial or total loss of the ability to articulate ideas or comprehend spoken or written language, as a result from damage to the brain caused by an injury or due to a disease.

Injury to Wernicke's area, results in a receptive aphasia, whereby the word production is normal, but the use of the words is defective. Injury to Wernicke's area causes, for the listener, a difficulty in translating and understanding the meaning of what someone has said to them. People with receptive aphasia will tend to string words together without apparent meaning.

Injury to the Broca area causes a motor aphasia resulting in slow, prolonged output of words, poor articulation, short sentences, which is known as an expressive aphasia.

Autistic individuals may have a nerve or implementation problem (hardware or software problem) in either Wernicke's area or the Broca area, and therefore exhibit difficulty with either understanding spoken language (a lesion in Wernicke's area) or difficulty in formulating meaningful spoken language (a lesion in the Broca area).

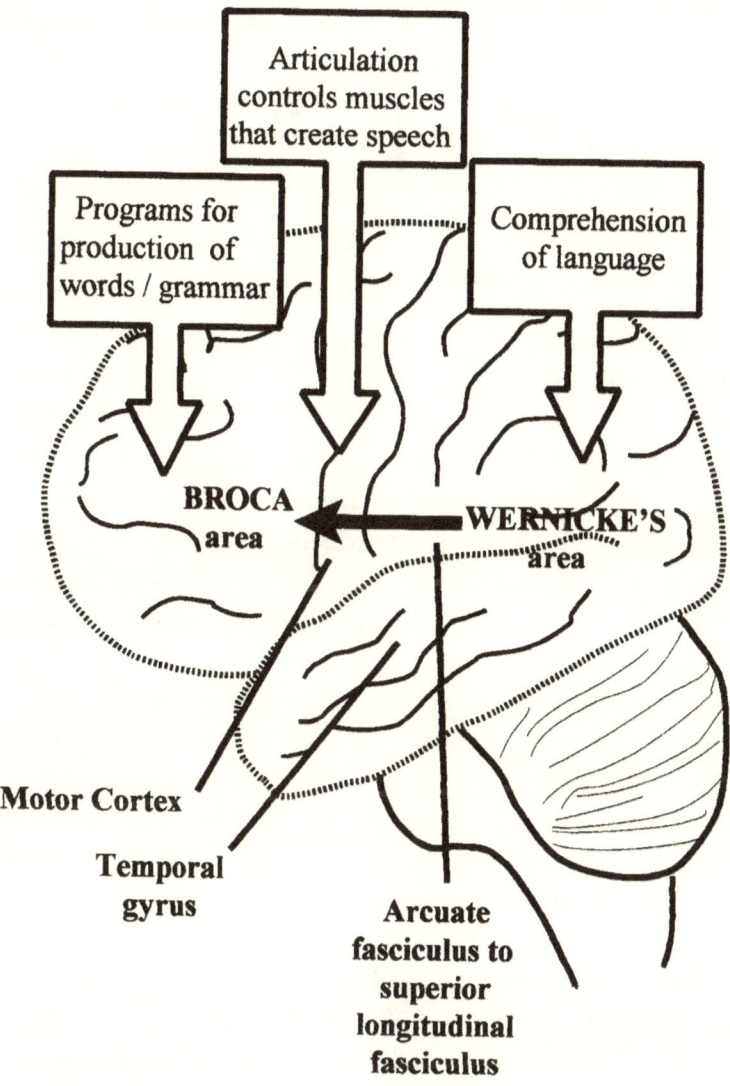

Figure 44.   Areas of the Human brain auditory interpretation and speech production (left side view).

25

SMELL IT LIKE IT IS

Crude sensors allow computers to sense particle matter in the air. Therefore, a computer's sense of smell may be imaginatively thought of as a building security computer connected to smoke detectors or carbon monoxide detectors or other type of detectors fashioned to trap and measure other potentially hazardous chemicals that might be present in the air. Recently, devices have been developed to allow a computer to release smells into the air, to provide a computer's user with an aromatic experience while interfacing with the computer. Our understanding of what smell is and how smells are detected, is very limited. Our capacity to effect a sense of smell in current computer technology is equally limited.

As humans, our sense of smell is attributed to 100 million olfactory fibers that enter the olfactory bulb in the bone of the skull near the nose. Olfactory nerves in the nose sense smell. Twenty-five thousand nerve axons terminate in a congregation of nerve endings referred to as a glomeruli. The glomeruli send nerve impulses to tufted cells, which number approximately 150,000, and mitral cells that number 50,000. The tuft cells and the mitral cells transmit nerve impulses regarding the sense of smell to the brain by way of the olfactory nerve.

The nerve impulses of the olfactory nerve traverse to the olfactory trigone, then split and send nerve fibers to structures called the medial olfactory stria, lateral olfactory stria, the anterior olfactory nucleus, the olfactory tubercle, the prepiriform cortex, the amygdala, the septal nuclei, and the hypothalamus.

The nerve receptors located in the nose are considered the first order neurons. These first order neurons are nerves that transmit information from the sensor in the nose to a data collection point, in this case termed the olfactory bulb. The tuft and mitral cells, in the olfactory bulb, are considered second order neurons (second in the line of information transfer to the brain). The second order neurons transfer information to another data collection point called the olfactory trigone. Third order neurons transmit the important information to the amygdala, then to the cerebral cortex for further processing and sensory awareness.

The anterior olfactory nucleus and the olfactory tubercle send efferent fibers back through the olfactory bulb to the tuft and mitral cells. The tuft cells send efferent signals to multiple glomeruli, while mitral cells send efferent signals to only one glomeruli. The action of the efferent cells is thought to be that of negative feedback to dampen the response of the receptor cells. The neurotransmitter Dopamine acts to down-regulate olfactory information, helping to damper signals regarding useless smell information.

The linkage of the olfactory nerve axons to brain structures such as the amygdala, the hypothalamus, and the prepiriform cortex may explain how the sense of smell is integrated into sexual behavior and feeding. Since the amygdala seems to assist in processing a human's sexual behavior, having olfactory nerve endings terminate in the amygdala links smell to a potential mate. Since target acquisition directed at locating and relocating a suitable mate is a primary function in an adult, gathering as much information and linking the information regarding the suitable mate is imperative. Collecting data on the smell of the individual and the smell of the location surrounding the individual, assists a suitor in relocating the potential mate.

Being able to integrate information regarding smell with a potential mate and being able to integrate smell with food or associate smell with danger, helps to secure the survival of the individual and survival of the species.

Temporal lobe injury, disease states, and tumors have been known to result in olfactory hallucinations. That is, injury or a stroke in the temporal lobe region of the brain, may cause a person to think that they are smelling something, that is actually is not in the air at the time the individual thinks they are sensing the smell. Wouldn't that be a drag?

26

GRAB IT LIKE IT IS

Your desktop computer doesn't have to worry about putting expressions on its face, or getting food to its mouth, or scratching its head, or walking down the street to the mailbox to see if any surprises were dropped off by the mailman. The desktop computer does not have to concern itself, or dedicate memory space, or CPU processing power to movement or balance. Well, the computer might have to open the DVD or CD drive when you push the proper button, but other than that, the computer lets you move the mouse.

Humans on the other hand, like most other animals on the planet, do have to worry about balance and movement. Where most animals surviving in the world have to move in some capacity because they still live under Mother Nature's harsh rule of *Eat or be Eaten*. Humans: how long could we survive without being able to find the remote for our television set, or fix a bowl of ice cream topped with nuts and chocolate syrup, or slam a bag of popcorn cornels in the microwave? How are these vital tasks accomplished you might ask?…Through a very intricate, well orchestrated system of sensors, balance regulators, brain processing power, movement strategy formulation and reformulation, and elaborate arrays of muscle motor commands. How else can we explain being able to drive our car, speak on the cell phone, chomp on a donut and listen to the car radio all at the same time?

All over the exterior wrapping of the human body, embedded in the skin of the torso and the limbs, are located tactile (touch) sensors, proprioception (position) sensors, temperature, vibration and pain sensors. Information gathered from temperature and pain sensors is routed from the legs, abdomen, torso, back and arms to the opposite side of the spinal cord, to the spinothalamic nerve tract. This nerve tract acts as a data bus to route fast pain and temperature information up the spinal cord to the brainstem to the thalamus and then the information is routed to the posterior (back) portion of the parietal lobe where the signals are processed. Information concerning tactile, vibration proprioception data gathered from sensors located on the legs, abdomen, torso, back and arms is routed through the Geniculate and Cuneate nerve tracts (from the same side, up the same side of the spinal cord) to the Gracile nucleus and the Cuneate nucleus located in the Medulla of the brainstem. The information is then transferred through the medial and lateral lemniscus that decussates (crosses over) to the opposite side of the brain. The tactile, vibration and proprioception data is routed to the Ventral posterolateral nucleus (one of at least eleven nuclei) in the Thalamus. The data is then routed to the posterior (back) portion of the parietal lobe on the opposite side of the body from where the sensors are located.

The frontal portion of the Human brain makes decisions as to what tasks it wishes the body to accomplish. These desires are relayed to the subconscious. The subconscious Human central processor, comprised in part of the Thalamus and the Lentiform nucleus, request movement strategies to be formulated by the frontal (forward) portion of the parietal lobes. The parietal lobe on the left side of the brain is responsible for muscle motor control of the limbs on the right side of the body. The parietal lobe on the right side of the brain is responsible for muscle motor control of the left side of the body. The forward portion of the parietal lobes coordinate their strategies with the sensory information gathered in the posterior portion of the parietal lobes (note: the motor movement and the sensory data collection center are separated by a depression that courses down the side of the brain known as the central sulcus), and the cerebellum (where balance and spatial orientation are processed), as well as any interrupt messages such as 'danger' or 'fear' warnings the brain may have stored in an interrupt memory file.

Once a strategy for an action has been devised, the subconscious Human central processor transmits the appropriate muscle commands down to the brainstem. Facial muscle motor commands are relayed to the muscles of the face by the cranial nerves. Neck, arm, torso, back and leg muscle motor commands are relayed to the muscles of the arms, chest, back and leg muscles by way of the pyramidal nerve tract that travels down the brainstem, then down the spinal cord. At each spinal level nerves interface with the pyramidal tract nerves, exit the spinal cord and traverse to muscles located in the neck, arms, torso, back or legs. The appropriate movement of an arm or leg, the chest or back is a constant task of real-time movement of the muscles of the body, requiring constant transmission of sensory data from the peripheral body back to the sensory processors located in the posterior (back) portion of the parietal lobes of the brain. The sensory processors in the posterior portion of the brain continue to update the muscle strategy command center in the frontal (forward) portion of the parietal lobes with new information. Thus, the muscle motor strategies can be continuously refined to accomplish the tasks of moving through the environment to satisfy the desires and needs generated by the frontal brain lobes.

Illustrated in Figure 45, is the Homunculus. The Homunculus is the division of the parietal lobes, both sensory input (posterior portion) and motor control (frontal portion), from top to bottom into portions of the body. The top of the parietal lobe represents the foot. Below the foot area is the leg, then the hip. Below the hip is the neck, then the shoulder, the forearm, then the hand and thumb. Below the thumb is the eye, nose, face, then the tongue. Below the tongue command center is the abdomen. The sensory processing and the motor command centers of each of these divisions are located in these areas.

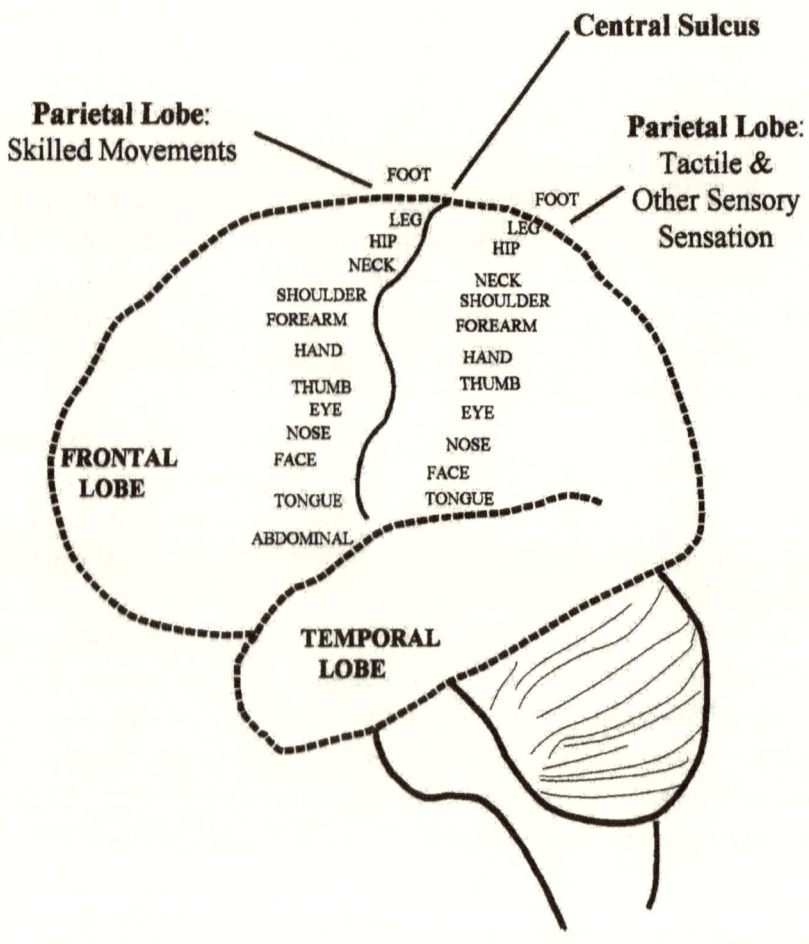

Central Sulcus

Parietal Lobe:
Skilled Movements

Parietal Lobe:
Tactile &
Other Sensory
Sensation

FOOT
LEG
HIP
NECK
SHOULDER
FOREARM
HAND
THUMB
EYE
NOSE
FACE
TONGUE

FOOT
LEG
HIP
NECK
SHOULDER
FOREARM
HAND
THUMB
EYE
NOSE
FACE
TONGUE

FRONTAL
LOBE

ABDOMINAL

TEMPORAL
LOBE

Figure 45. **The Homunculus** located in the Parietal lobe of the human brain (left side view).

Figure 46, illustrates the course of the pyramidal tract and the spinothalamic tracts. Figure 47, summaries the flow of information up

to the parietal lobes of the brain and the motor command signals down
the pyramidal nerve tracts from the parietal lobes to the extremities.

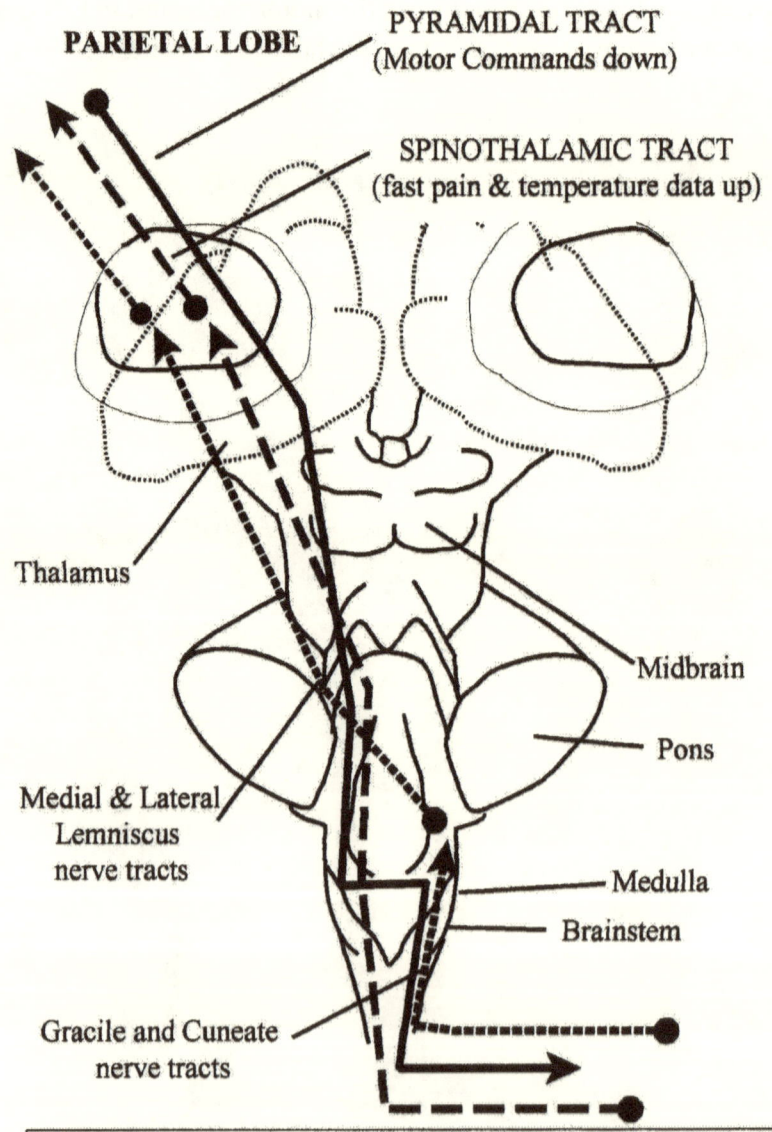

Figure 46. Pyramidal nerve tract sends motor command signals down to muscles, Spinothalamic tract sends pain and temperature data up to Parietal lobe for processing (front view).

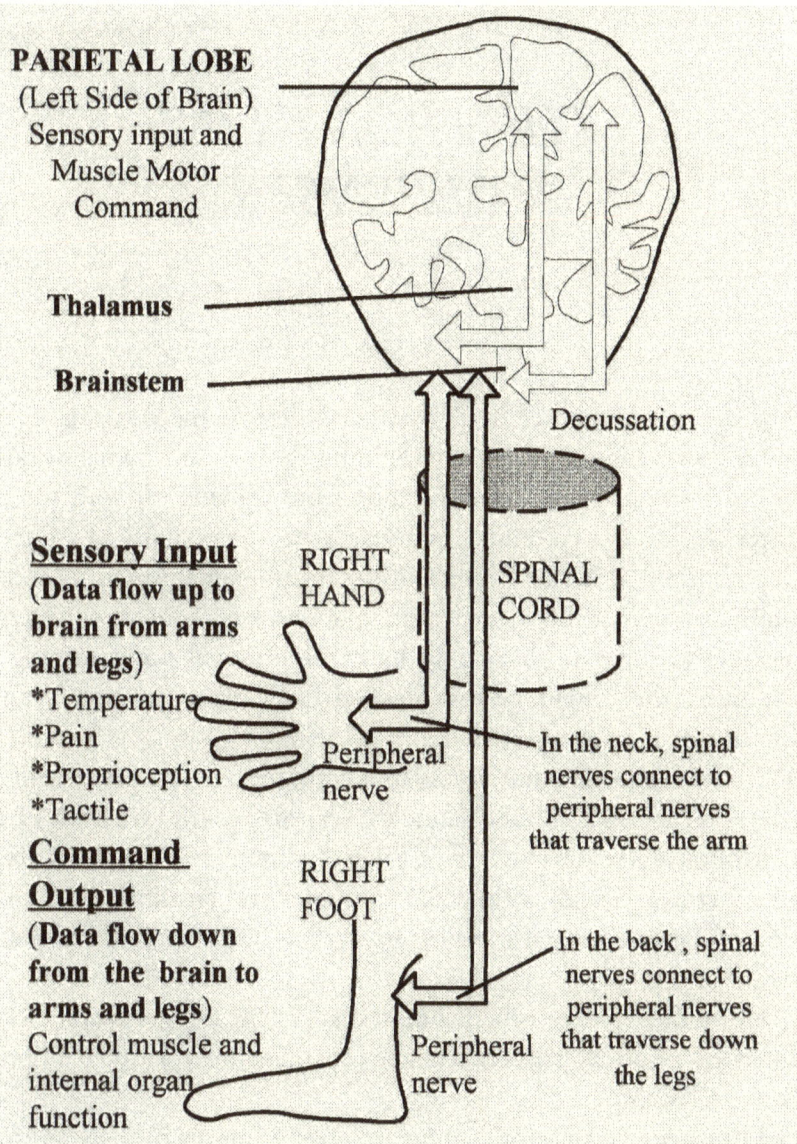

PARIETAL LOBE
(Left Side of Brain)
Sensory input and
Muscle Motor
Command

Thalamus

Brainstem

Decussation

SPINAL CORD

RIGHT HAND

Sensory Input
(Data flow up to
brain from arms
and legs)
*Temperature
*Pain
*Proprioception
*Tactile

Peripheral nerve

In the neck, spinal
nerves connect to
peripheral nerves
that traverse the arm

Command Output
(Data flow down
from the brain to
arms and legs)
Control muscle and
internal organ
function

RIGHT FOOT

Peripheral nerve

In the back, spinal
nerves connect to
peripheral nerves
that traverse down
the legs

Figure 47. How electric signals traverse between
the brain and the arms and legs in a person who is
right hand dominant.

27

THE FRONT LINE DEFENSES

The desktop computer may have a few defense mechanisms. A computer's owner might physically chain his desktop unit to his workspace in order to prevent a would-be thief from walking off with the computer. Inside the computer, passwords can be established to prevent unauthorized access to important files or programs. Antivirus and firewall software can be loaded onto a desktop computer to screen out destructive software that could harm the internal programming of a computer that might be downloaded from other computers on the Internet or local group nets.

If virus software gains access to a computer, it can use the infected computer to replicate and launch itself through the Internet or local network connections to infect other computers. Virus software can attack critical files in an infected computer and terminate useful function of the computer if the virus software is not identified and neutralized.

Computers infect other computers by spreading viruses over the Internet, by local network connections, and by corrupted floppy or corrupted compact discs.

Humans are constantly under attack by hordes of 'bad bugs' that circulate in the environment, most of which can only be seen with the aide of a microscope or in some cases, an electron microscope (for studying viruses). Humans spread infections to each other by coughing, sneezing, or even breathing on each other. Hand contact can spread infections either by directly touching another individual or by touching and contaminating drinking water, or food or the containers that water or food are transported in. Passing saliva by kissing is a common means of propagating infections. Sexual intercourse can spread many forms of infectious agents including AIDS, hepatitis, and gonorrhea. Sharing needles with other individuals places a person at very high risk of inoculating themselves with an infectious pathogen (generally a virus and/or bacteria).

The desktop computer may have virus software to screen out unwanted, malicious software. The human body has an elaborate form of protection against the many environmental pathogens. Pathogens come in many forms. The skin of the human body carries with it several forms of bacteria and fungi that are constantly ready to invade the body's deep tissue if the body's surface defenses ever fail.

Pathogens that actively would like to invade the human body include **bacteria, viruses, fungi, helminth, ameba,** and **protozoa.**

Bacteria are microscopic (need a microscope to see the darn little things) organisms with a cell wall and organelles (miniature organs) inside the cell wall, but no defined nuclear membrane. Due to the fact that there is no nuclear membrane, bacteria are considered more primitive than animal or plant cells. Most bacteria are single-celled organisms. Structurally, bacteria come in different shapes and therefore are classified in several different forms which include: spherical, known as *coccus*, rod-like known as *bacillus*, spiral or *Sprillum*, comma-shaped or *Vibrio*, and cork-screw shaped known a *spirochete*. Bacteria are responsible for numerous infections in man including Strep (short for Streptococcus) infections of the throat, pneumonias, urinary tract infections, sinus infections and many types of skin infections.

Viruses are minute particles, much smaller than bacteria, too small to be seen by a light microscope. Viruses consist only of a protein outer shell and a nucleic core that is comprised of either deoxyribonucleic acid (DNA) or ribonucleic acid (RNA), instructions for replicating additional virus. Viruses are only capable of replicating inside a living host cell. A virus will generally attack a living cell, penetrating its cell wall, then gain access to the cell's nucleus. Once inside the cell's nucleus, the virus will introduce its own DNA or RNA (instruction code) into the host cell's DNA. The virus programming instructions will force the host cell to discontinue normal physiologic function and, instead, cause the host cell to devote its resources to replicate the virus. When a significant amount of the virus has been replicated, the host cell's wall will rupture, releasing the copies of the virus into the surrounding tissue or environment. Viruses are responsible for the common cold, influenza, measles, mumps, chicken pox, herpes, AIDS, polio and rabies in man.

Fungi are single-cell plants that lack chlorophyll, a chemical that allows most plants the ability to conduct photosynthesis when exposed to sunlight. Common fungi are classified as yeast, rusts, molds and mushrooms. Fungi live as either saprophytes (lives on the dead tissue of animals or plants) or parasites (which live on a live host). Candida is a yeast that commonly infects women or people with immune deficiencies. Blastomycosis and histoplasmosis are fungi that infrequently infect people, some of which who also have immune system deficiencies.

Helminth refers to parasitic worms, which include flukes, tapeworms, and nematodes. Flukes are parasitic flatworms belonging to the group Trematoda. These flatworms have suckers that enable them to attach themselves to a liver, the lungs, the gut and to blood vessels, resulting in very serious diseases in humans. Tapeworms are long, thin ribbon-like flatworms that can infect and live in the intestinal tract of man. Nematodes are referred to as round worms and have an unsegmented cylindrical body. Nematodes include hookworms and pinworms that can infect the intestinal tract of man, and Filariae that can infect the lymphatic tissues.

Amoeba are microscopic, single-celled animals of a jelly-like consistency, that constantly change shape, moving and feeding in a damp environment, by the use of pseudopodium, or extensions of their body. Ameba such as Entamoeba cause disease in man.

Protozoa are single-celled free living animals. Common protozoa include Plasmodium which causes malaria, and Trypanosoma, which may cause sleeping sickness in humans.

To combat these many, and very aggressive microscopic villains the human body's defense system is comprised of white cells, internal organs that filter the blood, antibodies (proteins produced by B-cells to coat and kill bacteria) and complement (a natural part of the blood that, like antibodies, coat and kill bacteria). White cells are divided into T-Cells and B-Cells.

T-Cells act as roving police. When a T-Cell encounters an invading pathogen (possibly one of the above described villains), the T-Cell releases chemical signals that attract other similar T-Cells to the area. The T-Cell also releases chemical signals that flow through the blood stream that alerts the B-Cells to the identity of the invading pathogen. Select B-Cells that recognize the T-Cell signal, proliferate and start to produce protein molecules that coat the outer wall of the pathogen and assist in killing it. These proteins are referred to as antibodies.

T-Cells are divided into several different cell lines. Neutrophils are used to defend the body against bacteria (see Figure 48). Lymphocytes are used to attack and route out viruses. Eosinophils are used to defend against parasites such as amoeba, protozoa and worms. Macrophages are large scavenger white cells, that assist with the initial identification and mounting of an inflammatory response against an invading pathogen. Natural Killer cells are thought to seek out and destroy potential cancer cells.

The reason for bringing up the immune system is to identify that the human computer's defenses act, in an abstract form, like a desktop computer with preset command functions. The immune system stays surveillant for any sign of a breech of the body's perimeter defense system (mainly the skin on the outer body or lining an inner cavity). Once a successful breech of the perimeter defenses is identified, the body's immune system swings into action and gears up to repel the attack. Without any conscious effort on the individual's part, the appropriate line of T-Cells are sent into action, and the B-Cells produce antibodies that aide in the identification and eradication of the invading organism.

With the newer Antivirus and Internet firewall software being developed, the desktop computer is moving closer and closer toward the animated nature of the human computer's defense system.

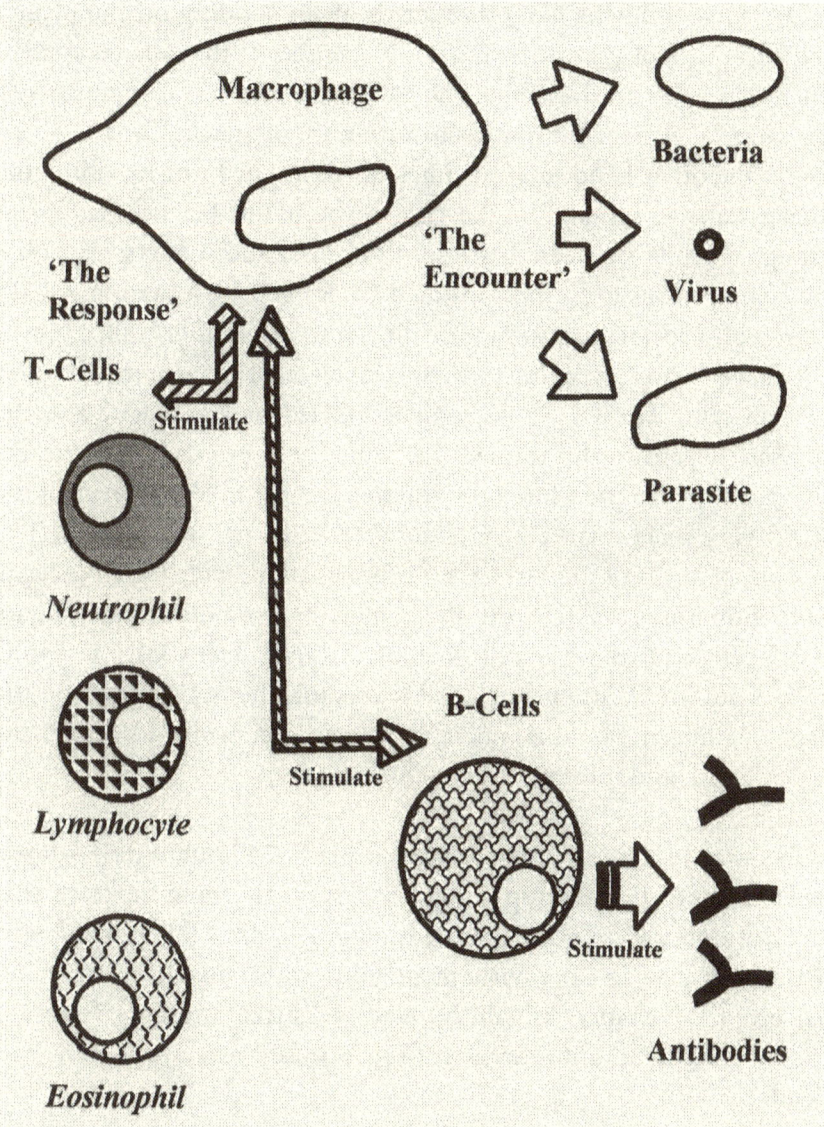

Figure 48. A macrophage encounters a bacteria, virus or parasite and stimulates a T-Cell and B-Cell response to the invading organism(s).

Of course, humans like other forms of life residing on the planet, take advantage of genetic diversity. In the case of the defense system, genetic diversity on the side of the pathogens means our immune system must be vigilant for numerous variations of attack. Genetic diversity on the other hand, may at times, work in our favor by aiding the human defense system. In the case of both the Plague that swept through Europe between 1347 to 1350, and the modern day HIV virus, populations in Europe possess a CCR5 genetic mutation. CCR5 is a receptor located on surface of the white cell's outer wall. Human cells have many receptor proteins that facilitate communication between cells. The CCR5 receptor is utilized by the Plague and the AIDS pathogens to slip through the white cell's exterior wall defenses and gain access to the white cell's interior cytoplasm. An absence of the CCR5 gene in Human DNA, alters the CCR5 cell wall protein. The presence of a pair of mutant CCR5 genes in an individual's DNA, may confer immunity to AIDS and the Plague. By not possessing an intact CCR5 gene, creates white cells that are missing intact CCR5 receptors on their surface. Without the CCR5 receptor, the AIDS virus and the Plague pathogen would not be able to gain access into its target, the helper T-cell, and infect it.

The world is an extremely dangerous place, 24 hours a day/7 days a week, for both the desktop computer (especially those desktops that stay continuously connected to the Internet), and for the human body, that simply lives in an environment filled with nasty little creatures, that given the chance, are always ready to strike, invading from air, land, water,...and unfortunately from possibly even your very best friend.

V. STRETCHING YOUR IMAGINATION EVEN FURTHER

Δ

INVESTIGATING THE PERFECT CRIME

"This is an open and shut case," stated a tall, lanky police lieutenant. The brash young officer was new to his job. He had been promoted to lieutenant due to his propensity to close difficult murder cases quickly.

"How do you figure George?" asked the heavy set, seasoned police captain.

Looking across the captain's cluttered desk, the lieutenant responded, "Well, the way I see it Cappy, the two doctors and their assistant meet the description of the perpetrators given by witnesses,…at least by their approximate height and weight. We found pieces of Washington's, Jefferson's and Ole Ben's masks in the office. We uncovered a stash of bills."

"So you're convinced these guys are responsible for all five Presidents bank heists on the east side?"

"Yep."

"What's the motive?" the Captain was quick to add, as he waved his beefy right hand around in the air, "George, tell me why would two prominent doctors and their assistant rob banks?"

"I don't know Jimmy…fill in the blank however you want."

"So you think these guys had a modern version of Dr. Jekyll and Mr. Hyde going?…Or they just did the robberies cause they wanted to?"

"Yah,…Something like that."

The captain insisted, "George, You gotta have a motive,…or you can't close the case."

"Cappy,…Maybe they had expensive girlfriends,…big gambling habits,…Or maybe they was just supporting bad drug habits."

"George, I think you still have work to do."

"Awe come on Cappy…" the lieutenant pleaded, "We got all three masks…and part of the loot…at least the part that didn't get burned up,…I say we call it quits." The lieutenant retorted, "Why spend any more of the taxpayer's money on this investigation?"

"Well, maybe your right." the captain acknowledged, "We could just say the doctors were supporting illicit habits." The captain paused then asked, "Anyone gonna argue with us?"

"No particular family in town for any of them."

"You sure?...I don't want to end up looking like some laughing stock."

"Both doc's parents are dead. They had no sibs," he added, "The assistant was a loner."

"They have friends?"

"People tells me the three worked day and night on some sort of heavy duty medical research project." The lieutenant continued, "A neighbor of the docs thought he overheard the two brothers talking about the Nobel Prize."

"So no one really knows these guys?"

"Not on a first name basis,...certainly not in the biblical sense,...a few barely beyond a last name basis."

The captain thought about it for a long moment then said, "Okay,...We'll make things easy on the robbery case." The senior officer added, "But what or who killed the three?"

Confident the lieutenant answered, "They nailed themselves."

"How do you figure?"

"The offices were filled with oxygen and nitrogen tanks," the lieutenant continued, "Some piece of computer equipment blew, igniting the volatile stuff and the whole place went ka-blewy!" The investigative officer threw his hands up in gesture.

"So these guys were just unlucky?...And that's how we caught them."

"Damn unlucky Cappy." The lieutenant said with a satisfied smirk stretched across his face.

The captain argued, "I don't buy it George."

Choking on his words, the lieutenant stammered, "What's there to buy Cappy?" To strengthen his position, he added, "They had a bunch of surgical and computer equipment stashed in them offices in close proximity to each other. Something got out of hand, one fatal spark and the place lit up like the Fourth of July."

"Any sign of arson?...Or better...murder?" asked the senior officer.

Staunchly, the lieutenant answered, "The fire chief's already combed the office space. He told me his report's going to say, he thinks the fire was accidental. There was no sign of a struggle. The corpses had no signs of trauma."

A satisfied smile finally stretched across the captain's pudgy face. The senior officer nodded his head in approval. He was pleased, George Rickless did seem like a good choice for a lieutenant...Maybe for once, his Department could close some of the old cases that had been hanging around the office collecting dust for too long.

The lanky lieutenant took his cue from the nod and lifted his body up out the chair he had nested in. As he reached for the handle of the glass door leading out of the captain's office, the senior officer asked, "Hey George,...What about the guy that committed suicide down on the street just before the building blew?"

"You mean the guy who tried to stop a six ton city bus with his bare hands and got mashed in the process?"

"Yah."

"What about him?"

"Well,...was he involved?"

"Involved in what?" a blank look stretched across the lieutenant's face.

"Involved in the bank robberies or the deaths of the doctors you idiot,...What do you think I'm asking?"

"Couldn't be Cappy."

"And why not?"

"He was dying of cancer."

"What kind?"

"Lung cancer, been struggling with it for eighteen years."

Spiteful, the captain muttered, "So he had nothing to lose."

"Yah, but the medical records we could locate, also say this guy was being worked up for some kind of serious brain disease…maybe cancer cells in the brain."

"So he has a donut, but had a bad batch of jelly?"

"Maybe more likely a fluffy French pastry with a lot of air holes…I don't know."

"Well no wonder he tried to play super hero with a city bus," the Captain remarked.

"He maybe never knew what smacked him."

There was moment of silence, then the captain asked, "What was he doing near the building?" He added, "Was he lost?"

The lieutenant retrieved his palm size notebook from the inside pocket of his forest green sport coat. He flipped to the page where he had recorded his notes on the bus accident victim. He read from his notes, "The guy's name was Nevin Thomas. A dentist's office was on the same floor as Dr. Stephens. Thomas had an appointment to see the dentist."

"Do we know if he ever saw the dentist?"

"The dentist's office records were burned beyond recognition."

"Did you think to ask the dentist?"

"Ask who?"

"The dentist, did he see Thomas that day as a patient?"

"No."

"No as in he wasn't seen, or no as in you didn't ask?"

"Didn't ask."

"Why not?"

"When the dentist's office was destroyed by the fire, he went on a vacation to Alaska. I guess he figured there was no reason to stick around town while his office was being rebuilt."

"Doesn't that sound fishy to you George?…excuse the pun."

"No,…I'd go on vacation if I were him."

"Have you tried to get in touch with him?"

"Nobody's been able to reach him. He's somewhere up there salmon fishing…with the polar bears."

"So this guy Thomas, had only a half-baked brain, he was dying from cancer and cared about his bad teeth."

"That seems 'bout to sum the vic up," replied the lieutenant.

"Not, I guess, a great candidate for being a bank robber or a killer."

"Not unless his shrink, or cancer doctor,"

The captain filled in, "Or his dentist's fees were extraordinarily high."

"Possibly,…But unlikely."

The captain finally conceded, "Okay George,…Close the case…Pin the two brothers and their assistant with the robberies."

"Right oh Cappy." The lieutenant gave the senior officer a sort of half salute.

"Make sure you put down they killed themselves in a freak accidental fire."

"You bet."

"I want your report on my desk by five o'clock."

"You'll have it." The lieutenant opened the door and impatiently marched out of the captain's office.

The captain leaned back into his leather chair, folding his beefy arms behind his head. It just didn't make sense that two doctors would take up a life of crime and rob banks, then get themselves blow up from their own medical equipment. But then he had seen stranger things before. People, sometimes,…just were not whom they seemed to be.

28

WHY I DON'T UNDERSTAND MY TEEN
&
WHY DOESN'T MY TEEN UNDERSTAND ME
-INTRODUCTION TO THE LIMBIC SYSTEM-

John Riz, an eighteen year old high school senior, came home late one Saturday night. In the driveway stood his grandfather's car. John guessed that Grandpa Riz was staying over the night in the guest room, which Grandpa often did.

John slowly pulled in the driveway, and parked the family car in behind his Grandpa's car. The teen threw the transmission into park, cut off the engine, and quickly extinguished the headlights.

The teen opened the car door on the driver's side and stepped out. Above, the sky was clear. A billion stars twinkle in the pre-dawn hour. A full moon high in the sky bathed the landscape with a gentle, dim light. John glanced down at his watch. The illuminated face of the timepiece read three in the morning. A cool breeze gently brushed across the teen's cheeks.

John looked up at the house. All was dark except for the porch light illuminating the front door. Before stepping up to the house, John shuffled around to the rear of the car and peered down at the back bumper. The smooth surface of the automobile's rear bumper was marred by a two-foot wide dent. John had to see it one more time. It was the result of a stupid decision on his part. He had backed up the car in a parking lot without physically checking behind the car before doing so, despite that being one of the rules his father had drilled into him during his driver's education. The car had struck an immovable concrete post standing only two and a half feet off the ground. Despite the yellow paint that covered the malicious post, John had not seen the post, till it was too late. A gut wrenching, metal crunching sound, and a wicked jolt to the car, had alerted John that something was terribly amiss.

Yellow paint highlighted the crevices of the bumper's dent. John was now certain, it wasn't a dream, the rear bumper on the family's car was indeed damaged. John couldn't imagine what the repair costs would be. Exhausted, the teen rubbed his eyes. He couldn't imagine what his father would say when *he* saw the dent in the bumper…Correct that, John couldn't imagine what his father wouldn't say. He was certain to face an hour-long lecture from his dad, during which he would assuredly get his butt chewed out a hundred times or more.

Emotionally spent, John sluggishly dragged his dejected body up to the front porch, unlocked the door and slipped into the shadows inside the house. He dared not turn on one of the interior house lights. He had no desire to begin catching the wrath of his father at three in the morning. John shut the door and flicked off the porch light.

At exactly seven-fifteen in the morning, a robust, angry cry swept through the house, causing the walls to vibrate. "John Thomas Riz…Get your butt out here!"

Mr. Riz, John's father, had spied the dent in the rear bumper of the family car while retrieving the Sunday newspaper from the driveway. By the sound of John's father's voice, he was not happy with what he saw.

John bolted up out of bed. He still wore the same clothes he had had on the night before. He had been too tired to disrobe. He bolted downstairs.

Standing before his father, he stuttered as he tried to offer an explanation.

Breathing fire like an angry dragon, John's father, without thinking, blurted out, "John your paying for this in full…And you're grounded for a month!…maybe more!"

Devastated, John pleaded, "But Dad, the Prom, it's only two weeks away!"

Smoking like a coal fired steam engine, Mr. Riz barked out, "Well you should have thought about that before you crashed the family car!" he continued in the same breath, "Son,…you can forget going to the PROM, unless it's going to be held in our living room,…and I don't think so!" Jack Riz, the father, snorted as he shook his head back and forth.

Now wait a minute, what if the above interaction played out like this:

As John slept, he was gently nudged awake. A concerned voice said, "John, John are you hurt at all?"

John rolled over in bed and stated, "No Dad, I'm okay."

Mr. Riz remarked, "I saw the damage to the car, I came up to see if you were okay."

"Yah, I'm okay Dad."

"Anyone get hurt?"

"No Dad."

Maintaining a calm voice the father asked, "How'd it happen?"

"You mean the dent in the car?"

"Yes, I mean the dent in the rear bumper of the car."

"I hit a post in the parking lot at the movie theater."

"Where you horsing around with the car?"

"No Dad."

"We're you doing drugs or drinking?"

"No Dad."

"Why didn't you see the post?"

John confessed, "Dad,…I know I should have, but…I didn't check behind the car before backing up."

"Oh. You know that's not what I taught you."

"I know Dad," feeling a lump the size of a baseball sitting in the middle of his throat, John added, "I promise, I'll pay for the damages Dad."

Mr. Riz nodded his head as he stated, "Ya sure,…We'll talk about that later."

"Okay Dad."

"You sure you're not hurt?"

"Dad,…I'm okay."

"What time did you get in last night?"

"Three."

Mr. Riz shuffled up to the bed and pulled the covers over his son as he stated, "Get some more sleep,…We'll talk about this later."

Sheepishly John asked, "Dad…am I grounded?"

The father straightened up, paused for a moment, then answered, "Naw, you can't be,…The Prom's this weekend." John's dad added, "Like I said, we'll talk about it later. I think you're going to find paying for the repairs is going to be more than enough punishment."

Regarding the above two scenarios, the first is somewhat realistic and the second seems more like a fantasy…But should it be that way?

Introduction to the Limbic System

The word *limbic* means 'border' or 'margin'. The limbic system is thought to be the portion of the brain responsible for learning and for processing emotions.

Depending upon the reference one researches, the recognized components of the limbic system may vary. In general, the limbic system is thought to be comprised of the parts of the brain that include the cingulate and parahippocampus gyri, the amygdaloid complex, the septal region, preoptic area, hypothalamus, anterior thalamus, habenula, and the central midbrain tegmentum (see Figure 49). There are no specific defining borders that encompass the limbic system. The components have been arranged by neuroanatomists. Thomas Willis in 1664, pictured this region of the brain and gave it the name limbus.

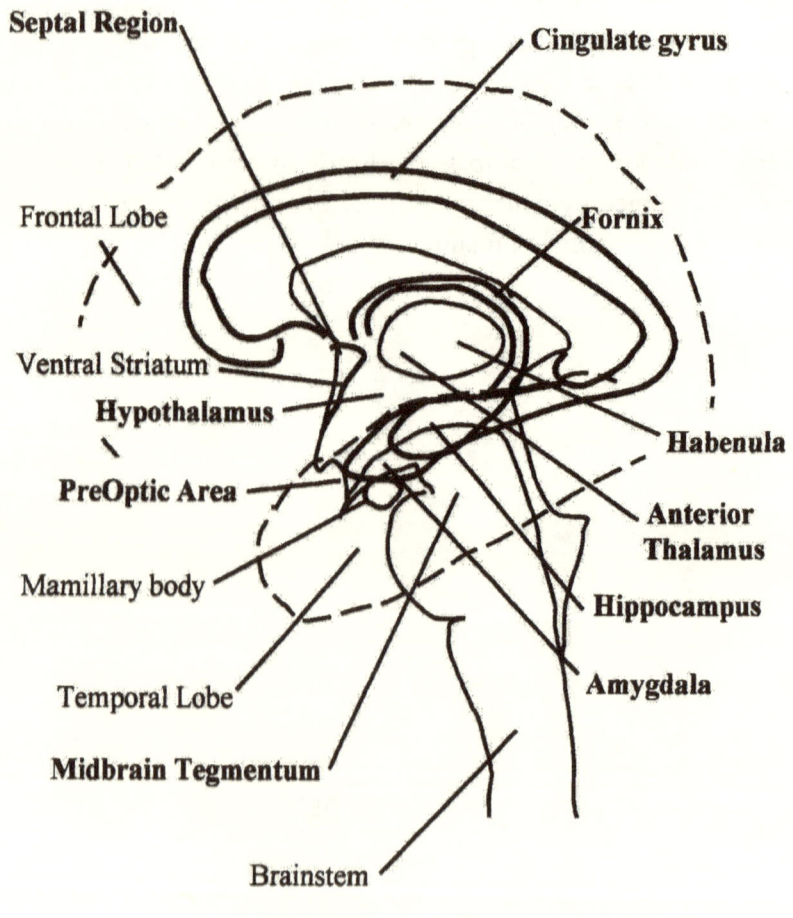

Figure 49. Major components of the
Limbic System (left side view).

The Limbic system is thought to be responsible for processing both
emotions and is important in learning (see Figure 50). Emotions are
considered to be any strong feeling state that is associated with bodily
changes and usually leads to an impulsive action or a certain behavior.
Emotions include fear, anger, love, hate. The Papez circuit, which is

comprised of the fornix, hippocampus formation, and the hypothalamus appears to be important in the learning process in humans.

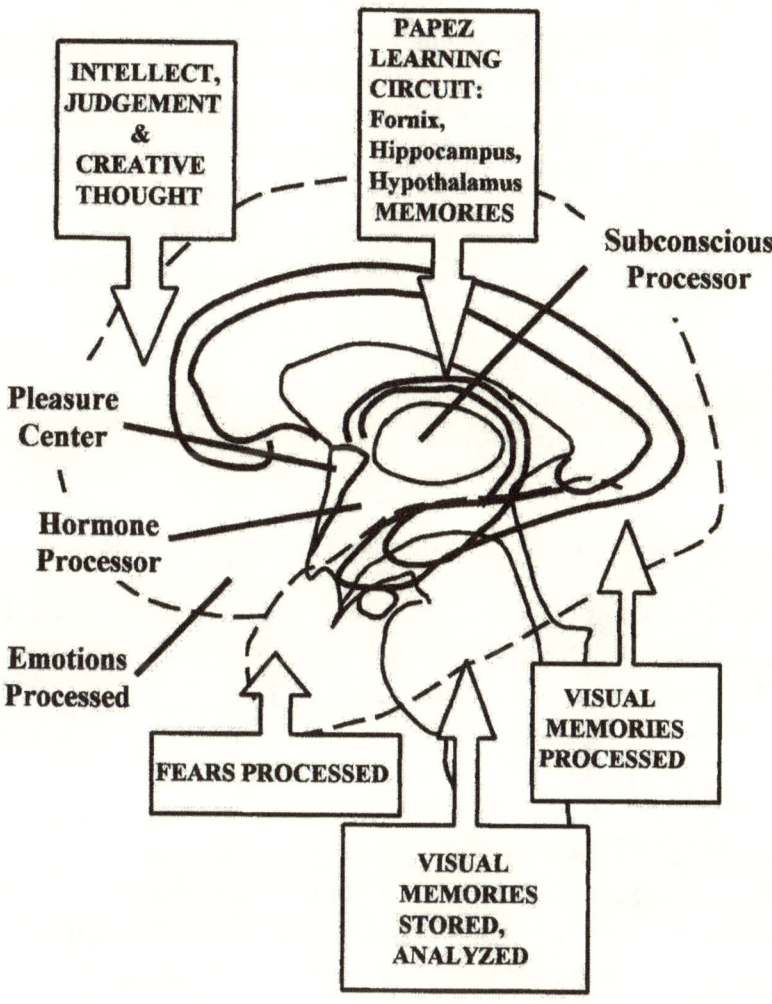

Figure 50. Major areas of function that comprise the Limbic System (left side view).

The teenage years become especially trying for many individuals as they struggle to understand their emotions. This process of understanding the strong feelings stirring inside their brain, and the bodily functions that accompany the feelings, is complicated by the bursts of hormones that occur in the young teens. Initially, the resultant behavior related to the emotions is that of outbursts and impulsiveness. The impulsiveness may be tempered by what has already been learned as socially correct behavior. As time marches on, and one learns the consequences of impulsive behavior, one attenuates their reaction to the emotions that stir inside their brain. Maturity is often judged as one's ability to experience the entire spectrum of emotions, but control the behavior generated by the emotions to produce constructive action. Immaturity is often a label applied to individuals who act out in a damaging or self-destructive action related to emotions they experience that they either can't control or do not choose to control.

There is another way to act out the above described scenario:

Mr. Riz shoved the side door open and stormed into the house. His mind was full of all kinds of profanities, but he kept them to himself. Mr. Riz threw the paper he had picked up from the driveway down onto the floor. The intent was to announce his presence in the house, and with the force used, the paper slapped the floor with a loud whip-like sound. The intention of creating the noise, was to alert all persons in the house of his displeasure. Without saying a word, Mr. Riz marched up the stairs to the second floor. Each step was delivered as if the man's legs weighed two hundred pounds each. The frame of the house resonated with each step. Reaching his son's bedroom, the father flung open the door.

John lay sprawled out in bed. Bed sheets covered the boy from head to toe. He was in such a deep sleep, he had not heard the raging tornado that whirled its way toward his room.

Spying his son still sleeping in bed after eight o'clock in the morning simply fueled the fire of the already angry father. Taking a deep breath, the man of the house filled his lungs to capacity. Then letting out the air like a blast of wind from a foundry furnace, he screamed out, "John! What the hell did you do to the car!"

The half conscious son bolted up in bed. The sheets fell to the sides. Naked from the waist up, John shook his head from side to side.

Grandfather Riz strolled into the bedroom behind *his* son. The grandfather stood silently and watched his son chew out the boy. When Mr. Riz was done, in a calm voice, the grandfather said to *his* son, "Now Jack,...Don't you think you're being a little tough on John." Jack, the father, turned to face *his* father as he nearly choked on his own saliva.

The last scenario plays out in this way:

The teenager is working with a limbic system that is still in a rudimentary form. The teenager's limbic system is still struggling to learn *how to learn*. The teen is not terribly skilled in assimilating information, nor is the teen skilled in making quick and important decisions. The teen has acquired numerous rules and regulations from parents, teachers and other mentors, but has not attached much physical meaning, nor does the teen have a real sense of consequence associated with the violation of these rules and regulations. The teenager's limbic system is at the mercy of surges of impulsiveness, in part due to the emotional center in the frontal cortex and the output of the hypothalamus and the pituitary gland. Still maturing, it may be difficult for many to understand the sporadic processing power of the brain. Hormones fluctuate in the teenage years. The effects on the body of sudden bursts of adrenaline and testosterone in a young male, and bursts of adrenaline and cyclic elevations of estrogen and progesterone in a young woman, may be hard to understand, interpret and control. In the young person, the experience that comes with age, is not present to temper these surging, and unpredictable emotional and hormonal factors.

In the father, the limbic system is much more developed and mature. The manner by which one learns new things is understood and exercised by the individual; the act of learning is set into a pattern by which the individual is comfortable. Assimilation of facts and making decisions on real life issues has become second nature. The father has learned to be successful by making appropriate decisions and seeing such decisions flourish. The father has also witnessed, first hand, the consequences of bad decisions. In fact, there may have been many bad decisions made along the pathway of one's maturity that end up leading to making more successful decisions later in life.

A 'heated reaction' by a parent to a teen's apparent 'irresponsible behavior' is probably the limbic system assimilating the facts and the burst of adrenaline that occurs when the parent realizes what the consequences of the teen's act or failure to act, might create. In the case of a car accident caused by a teen, consequences might include loss of time and money to repair the damage to the vehicle, injury to the teen that may interrupt the routine of daily life, or injury to another party that might end up generating legal action against the teen and the parents. The parent reacts, possibly over-reacts to a situation, because the parent is assimilating the facts of the occurrence, the fact that a rule or policy was broken or not followed (thus a bad decision was made on the part of the teenager), the potential consequences of the occurrence, the frustration over the apparent inability of the teenager to understand, and finally a burst of adrenaline or other hormone surge created the emotional attachment the parent has with their teenager.

The grandfather steps into the picture. The grandparent's limbic system is well experienced in learning, assimilating facts and making meaningful decisions after having lived long enough to have had to have made a number of difficult and meaningful decisions. The grandparent also has survived numerous meaningful life events and has a long history of experiences from which to draw upon when confronting most situations. In the case of a car accident, the grandparent has most likely lived through his own accidents, as well as one or more accidents that involved his children and friends. The teen's accident, is one of many, in a string of accidents and other life perils that has occurred in the grandparent's life. The grandparent's response is also tempered by attenuated bursts of adrenaline that the father and the teen are prone to experiencing.

The grandfather states, "Well Jack, I seemed to recall late one night you dragged yer butt into the house after having smashed the front bumper of the family car into a fence post…a rather *large* fence post you claimed *you* never saw."

The father growled as he instantly recalled the details of that night's events, "Ya, and you wanted to tar and feather my hide that night."

The grandfather confessed, "Well, and I have to admit, I was pretty harsh on you Jack…Too harsh for your own good."

Sternly Jack snipped, "Times haven't changed."

The grandfather piped up, "Yes, they have Jack. We didn't have things like 'time-out' when your mother and I was rearing kids." He added, "Taking time to catch your breath,…and think rationally before committing yourself to an action you really don't want to impose upon your son…*really* is a good idea."

Jack opened his mouth to say something in rebuttal, but realized it was useless. His father was right. The world was changing and impulsive behavior, no matter how right it might feel at the moment, generally lead to creating greater problems that would occur if the time were taken to plan out the response and calculate the repercussions.

Still, not feeling comfortable about tempering his authority, the father turned to his teenager and stated, "John, we will talk about this after you have showered and had breakfast." The father turned, and marched out of the room.

The limbic system is a very complex, very important part of the brain, and vital to facilitating the learning process. Like other, more obvious parts of the body, the Limbic system needs time to develop,…it unfortunately, does not come as a plug-and-play computer feature. The Limbic system needs to mature, so as to be able to assist the consciousness to make the right (successful) decisions to life' complex problems.

Finding Your Destiny

Another great conflict between parent and child occurs when a child chooses a career. Generally, the scenario plays out that the parent has spent their whole life invested in a career or building a business, and the child demonstrates *no interest* in following the same path as the parent. Such instances include the parent who is a physician, and the child aspires to be a lawyer. The parent that is an engineer and the child who dreams of being a physician. The parent that is a baker, and a child that wishes to be a dancer. The parent that is a police officer and the child that wishes to be the used-car salesman. Throughout all walks of life, the same conflict occurs time and time again. The son or daughter demonstrates no desire to choose the career or business to which the parent has devoted their entire life.

Why?…Why?…Why?…One might ask.

This conflict can be very frustrating to both parties, and even split relationships for a lifetime.

But if we think of this in terms of what is good for the community, *diversity* is what is important for the survival of society.

Consider a town comprised of 250 people. Let's say the town's baker's wife gave birth to ten boys. If each of the ten boys became a baker like their father, then the town would end up with 11 bakeries. If each of the boys' wives gave birth to ten boys, and each of these boys became bakers like their fathers, then by the second generation, the small town would end up with 111 bakeries. Society could not support such narrow aspirations of succeeding generations and would not survive. In addition, if the town's only baker died before he could have children, or train his children to be bakers, the town might be left with no baker.

By causal observation, most communities around the United States and in other countries, are comprised of similar components. Most communities have government, have police, teachers, fire fighters, bakers, lawyers, doctors, garbage collectors, plumbers, salespeople, mechanics, factory workers, etc. How does this happen? How is it that the east coast of the United States doesn't have all of the doctors and the west coast doesn't have all of the lawyers, or vice versa?

When one opens a word processing software to write a letter, a resume, labels, or a certificate one can choose from a list of already typed forms. These predesigned forms are called *templates*. In some cases, the detail of the template may be adequate enough such that the user of the software only has to type in a few specific details, and the entire letter or certificate can be printed.

Extrapolating the concept of a template to humans, in order for society to survive, society must have numerous jobs fulfilled. In each community there must be politicians, police, teachers, fire fighters, doctors, garbage collectors, plumbers, entertainers, salespeople, mechanics, factory workers, etc. It would make sense, then, that the human brain would have templates. Such templates would be files stored in the human brain that would open up in some people, while a person was maturing, that would provide a sense and bestow a measure of talent to an individual, such that the individual would aspire to fulfill the role dictated by the template.

So that if one's father is a physician and the father enjoys being a physician, well the son's template might lead the son to being a use-car salesman. If a mother is a physicist, her daughter might have aspirations of becoming a physicist, or the daughter might wish to become a nurse, because inside her brain she has the template of a nurse, that is preset program files that coax her to wanting to become a nurse. Templates may be overridden, but this may cause a measure of frustration to the individual, who feels compelled to follow the aspirations, directives of *their* template.

I have heard of a few instances amongst my colleagues where some became physicians because that was the desire of their parents. To appease the parents, they struggled through medical school and became a physician. Once the parents died, the individuals gave up the practice of medicine to become what *they* wanted…a chef, a postman, a computer programmer.

Templates, predesigned formats stored in the human brain, are vital to creating and maintaining the fabric of society. We are creatures of a higher order, and thus, our programming is very sophisticated. For Humans to survive and flourish in the harsh elements of the world around us, our programming required: (1) design of a very complex body, (2) the design of a very complex ecosystem that surrounds and supports us, and (3) the design of society to act as a platform upon which great accomplishments could be dreamed of, designed, nurtured, and finally achieved.

As hard as it is to face, everyone has their own template which dictates how one might contribute to society. Encouraging and supporting one to follow their template, and thus fulfill their destiny, is one of the greatest gifts we could give someone we love.

29

TARGET ACQUISITION
&
THE KISS

The body has a number of needs. It requires air, water, and food. In the wilderness survival classes I have attended, the instructors have mentioned the brain can survive only five minutes, at normal body temperature, without oxygen; the body can survive three days without water; and, in general, three weeks without food. Beyond air, water and food, the adult needs to find a sexual partner if one expects to procreate the species. The human cannot physical generate any of the above-mentioned items. The human must acquire theses items in order to survive, or in the case of procreation, so the species can survive, which at times may be a stronger driving force than that of individual survival.

The human computer must be able to recognize these targets and pursue acquiring these targets when the opportunity exists and the target comes into the individual's field of view (see Figure 51).

Inside the adult human there are two memory sites where the human stores visual images of targets of acquisition. One register is a revolving target acquisition register. Depending upon the priority given to a need or an item, the brain has the target acquisition register filled with that item. In the case of primitive man, let's say he is spending the afternoon searching for food. In his mind he is hunting for deer. The target of acquisition at that time is a deer and the target acquisition register has an image of a deer in the register so that the man knows what he is looking for. If while he is hunting in the woods, he stumbles onto a plant with white berries, he will look at them. The human brain will investigate the visual find by searching through the target acquisition registers which hold images of appropriate foods. The images of blue berries and red berries may come up, but white berries should not. Since the target acquisition register does not find a match of the white berries to an image of the food in the human's brain, the human will continue his search for the primary target of acquisition,—the deer. If the day is hot, and the man becomes exceedingly thirsty, then his priority of target acquisition may change. Later on in the day, the hunter may change the target of acquisition, from searching for a deer, to that of searching for water, if satisfying the requirement of thirst becomes a higher priority than satisfying the requirement of hunger.

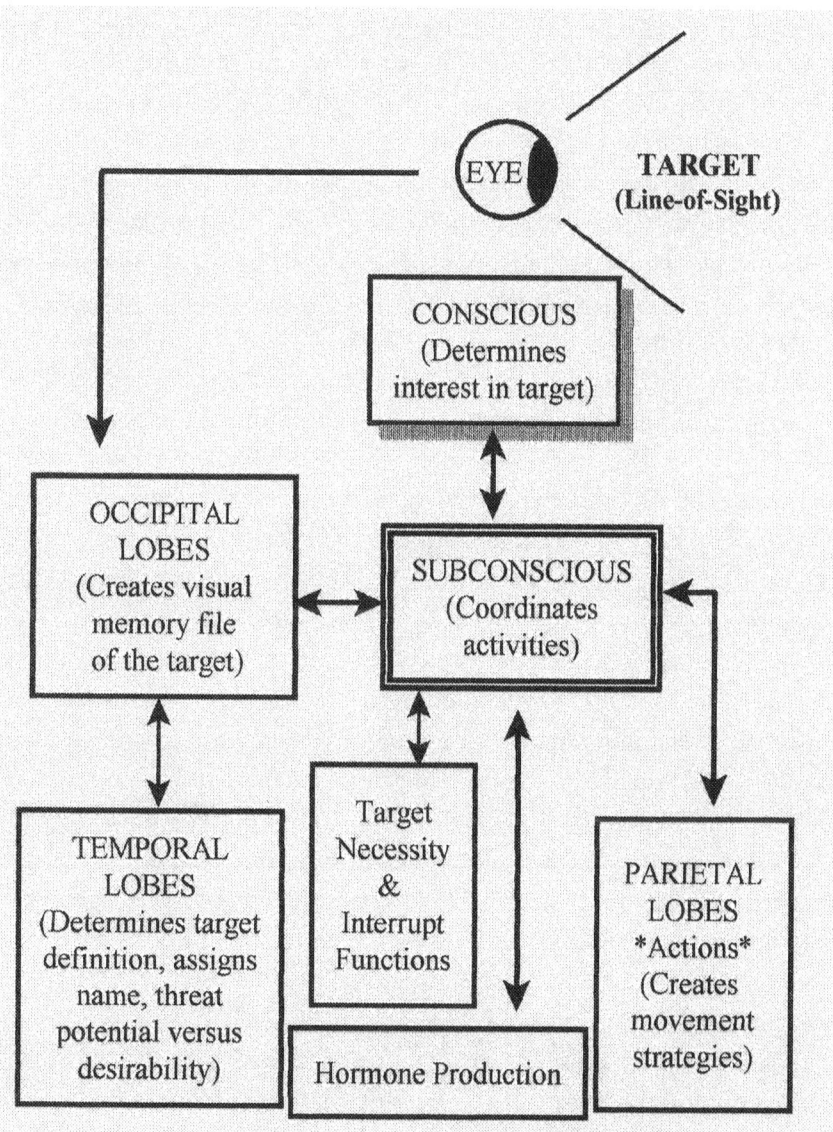

Figure 51. Diagram of the coordinated efforts of the brain when the eyes spy a target in their visual field.

A second target acquisition register is present in the adult human. The second register stores the image of what the opposite sex is suppose to look like. That is, the ideal image or silhouette of a sexually mature member of the opposite sex, is an image we store in our brain and is used to acquire a sexual partner. The size and dimensions of the member of the opposite sex, that an individual is seeking, may vary from person to person. However, the overall image of the sexually mature member of the opposite sex is somewhat standard amongst heterosexual adults.

So when the primitive human is searching through the woods for a deer, and stumbles upon a sexually mature member of the opposite sex, the search for food may become down-graded as the hunter analyzes the image of the member of the opposite sex. First, the hunter will visually scan the individual for physical evidence of sexual maturity. In the case of a heterosexual man analyzing a woman, the man's eyes attempt to match up the curves of the woman's torso and waist, the size of a woman's breasts, the presence of pubic hair, and how flat the abdomen is (a large lower abdomen may suggest the woman has already successfully mated with another male). In the case of the heterosexual woman analyzing the man, the woman scans the man for the presence of body hair, the size of his muscles and the size of the buttocks. If the individual the hunter is scanning turns out to be sexually mature, then the hunter looks closer to see how close the image of the individual matches the image of the ideal mate the hunter has stored in his head. If the sexually mature individual does not look at all like the image stored in the hunter's head, then the hunter may bypass any interaction with the individual other than to recognize the individual exists and go on his way. If the sexually mature individual looks close enough to the image stored in the hunter's head, then the subject is a prospective target. The hunter may start to pursue the individual for the purposes of being a sexual mate. The hunter's brain then evokes the function of his or her logic unit in order to generate the best strategy to pursue the individual. The logic unit helps the hunter recognize the sequence of steps required to identify if the target individual is a suitable mate and if the target individual is a willing sexual partner.

The regulation of target acquisition directed towards finding a sexual mate may be coordinated by the limbic system of the brain. The limbic system formally encompasses the parts of the brain that include the amygdala, hippocampus, septum, cingulate gyrus, cingulate cortex, hypothalamus, the epithalamus, anterior thalamus, mamillary bodies, and fornix. Hormones tend to enhance the sexual drive.

In men, testosterone is the hormone responsible for the development of the physical male sexual characteristics, and for helping drive the sexual interest in a man.

In women, estrogen is the primary hormone responsible for the development of the physical female sexual characteristic. Estrogen is thought to help drive the sexual interest in a woman, but testosterone levels may also play a role, at times, in the sexual appetite of the woman.

The point of discussing the sex hormones is that there is most likely an up-regulation in a search to satisfy finding a match to the image stored in the sexual target acquisition register when the sexual hormones are circulating in greater amounts in the body. The testosterone in a man is generated predominantly in the testes. The estrogen and progesterone present in a woman is predominantly generated in the ovaries by a maturing follicular cyst. A man may be stimulated towards sexual behavior based on his serum (blood) level of testosterone, which may be tied to the last time he ejaculated. If a substantial time has lapsed since the man last ejaculated (released his sperm) the testosterone levels may be high and the drive to find a sexually mature partner may become increasingly up-regulated. In addition, the higher the levels of testosterone circulating in the blood stream may down-regulate the requirement to find an exact match to the image stored in the target acquisition register. That is, if it has been a long time since the man has ejaculated and the man's testosterone levels are running high, he may not be very choosy in his selection of a sexual partner if the availability of sexually mature women is limited. The woman's desire for sexual behavior may be based on the serum level of estrogen and to some extent how much testosterone the woman's body generates (sexual hormone production is shared in part by both sexes) and the point in her cycle she is in at any given time. The closer towards mid-cycle the woman is, the more sexually aroused the woman might be.

Note, social interrupts and logic may dampen the above behavior, but the underlying mechanism remains present.

Once the target, the subject of sexual interest has been located, the next step is to secure the relationship. First, eyes generally make contact and both individuals take interest in what they see. But kissing acts to secure a sexual relationship. The act of kissing demonstrates to both partners that they are trusting of each other. Kissing acts to show there is an interest in pursuing a sexual relationship. Kissing also triggers specific responses by the brain. When the lips meet in two sexually mature adults, the amygdala, a nerve center located in the forefront of the brain, becomes stimulated. The amygdala (part of the Papez circuit) stimulates emotional memory. Stimulating emotional type memory ensures that the individual will remember whom it is they are kissing and imprints certain important features of the individual in long term memory files to insure that they can relocate this individual with whom the kiss had been shared. In the pituitary gland at the center of the brain, vasopressin is released in the male and oxytocin is released in women. Vasopressin and oxytocin are hormones that act as stimulators. In a similar bonding relationship, when a woman breast-feeds her newborn, oxytocin is released in response to the baby suckling on the mother's nipple. In adult women, oxytocin acts as an important neurotransmitter to help a woman secure a long-lasting relationship. In heterosexual men, vasopressin acts as a neurotransmitter and stimulates bonding with a woman.

The kiss shared between lips given in a purely voluntary manner between two sexually mature adults, acts as an unspoken contract that a sexual encounter is not only possible, but highly likely, if not inevitable.

So the eyes, in combination with the senses, search the population of humans for a prospective sexual partner comparing what is seen to a target file stored in a memory file in the brain. Men generally have an image of the ideal woman stored in their target acquisition file in their brain and women have the image of a man stored in their target acquisition file. In some individuals that externally are men, they may have a male target acquisition file stored in their brain and some women may have a woman target acquisition file stored in their brain.

Once a potential sexual partner is located by the eyes, the individual is pursued at a level of how close they match the target acquisition file in appearance and voice pattern. The closer the match, the stronger the effort to pursue the subject.

Pursuit of the target is inhibited by a series of potential interrupts. Interrupts generated within the brain of either subject may dispel the possibility of a relationship, either immediately or anywhere during the length of the relationship. Interrupts may be generated by such conditions as social status, religious obligations, family responsibilities, job requirements, or the fact that the potential sexual partner does not have in their brain a target acquisition file that meets the appearance of the potential partner that is pursuing them.

Pursuit of the potential partner may be enhanced if the individual pursuing the potential partner has attributes that may be attractive to the potential partner. We all recognize these attributes as social status, fame, money, an abundance of worldly goods, and a quick wit or humor. Attractive attributes may influence the one being pursued to be not so selective in their pick of a sexual partner when their brain reviews the contents of their target acquisition memory file.

When two individuals become sexually attracted to one another, there are almost always potential interrupts that may detour the relationship. If the interrupts, whether they be social, religious or monetary in origin, are strong enough, the relationship will be dispelled. If the attraction shared by the two individuals is stronger than the physical or mental barriers generated by the potential interrupts, the relationship will move forward. The individuals close ranks, come within inches of each other, and then share a kiss either secretly or in public depending upon the level of social interrupts that are present in their environment.

The intimacy created by the kiss secures the relationship between the two individuals and strongly suggests a sexual relationship is expected, unless circumstances beyond the control of the two individuals occurs. The kiss is truly the keys to our heart, or possibly the secret lock-and-key code to our master programming.

The Danger of Charisma

The dictionary definition of charisma is *an exceptional ability to secure other people's devotion or loyalty*. Charisma is generally believed to be the ability for an individual to appeal to the likings of a mass number of people. We see such positive appeal play out around us on a daily basis. Individuals that exhibit this mysterious quality of *charisma* lead businesses, are successful in politics, act as spokespersons for advertising campaigns, entertain us in the music and movie media.

Charisma has been often referred to as an animal form of magnetism. The attractive force by which charisma works is as invisible and mysterious as the magnetic properties of gravity, that holds all objects to the Earth's surface.

The quality of charisma is cherished, revered, idolized, highly romanticized. Charisma is hard to explain and is best described as *you know it, when you see it* or *you know what you like, when you see it*.

It would seem that something, some sensor, or some program subroutine, must be active in our brains that recognize certain physical or audio qualities in certain people. These individuals somehow stimulate the reward or pleasure centers in the human brain, and downgrade our suspicious/defensive mechanisms and stimulate us to follow the behavior commands that such individuals convey.

On the scale of selling us on the idea of purchasing an advertised good, or enjoying a form of entertainment, much may not be at risk-other than spending the cash we have in our own pocket. In terms of generating conflict or encouraging violence on a mass scale, charisma can lead to disastrous, irreparable damage to an individual or society in general.

Understanding the programming features that create the illusion of charisma, is a key element in learning how the brain functions. One very important reason to read this book, is to recognize that the human brain functions like a computer. Certain stimuli will cause the human brain to react in a predictable manner.

Magicians entertain us based on being able to depend on our brain's predictability. Illusions, referred to as magic, are created by carefully calculating our response to their misdirection of our attention. In effect, a magician blinds his audience to what he or she is really doing, by directing our attention away from the physical properties of the trick by using lighting, mirrors, smoke, flash, you name it, they use it. We enjoy being tricked, we savor the thought of magic…we circum to believing in the illusions, despite how impractical they seem to be.

Not being aware that our brains can be tricked, that our brains do react in predictable manners, leave *us*, the individual, and *us* as a society, open to exploitation. Being aware of that pre-programmed responses exist in *us*, allows one to possibly recognize, when one is being exploited and retain the capacity to make intelligent choices and remain an *individual*, despite how the masses may be swayed. It is not an issue of making a choice that is a <u>danger</u>; it is the *inability* to make one's choice for oneself that is the danger.

Now let's take the *charisma* issue to the furthest extreme.

Consider humans encountering a hostile alien race. If such an alien race understood the mechanisms of our sexual programming and our reward programming better than we understand it, they could use those features of our programming to gain influence into our culture.

Simple fact, since most men claim they do not understand women—If an extraterrestrial race could decipher and utilize the basic programming of women, and were able to manipulate the women on the planet to gain their confidence and allegiance, the entire human race would most likely fall prey to such an alien species. On the other hand, if a hostile alien race showed up in the form of voluptuous, sexually enticing young maidens, the planet's armed male defense system would most likely collapse without a shot.

Let's consider for a moment, that an extraterrestrial race shows up on our doorstep, or that we venture out to the stars and cross paths with an alien race. If such an alien race were not as intelligent as us, as a consequence of the alien encounter, we may concern ourselves only with what diseases they might bestow upon us, that we have no natural defense against. An aggressive pathogen hanging on the coattails of any alien could inflict widespread illness if the human body had no means of defending itself against the pathogen. Countless thousands of Indians in the Americas died from small pox and other diseases carried by the early settlers of the Americas. The Indians' immune systems had not previously encountered these diseases, and therefore, had no experience in staving off and eradicating the settler's pathogens. Unfortunately, early settlers of the Americas killed many more American Indians than did the weapons they carried. Any blending of cultures, is accompanied by an exchange of microorganisms.

On the other hand, if humans encounter an alien race that is more intelligent than we are, they may understand our brain function better than we understand it ourselves. A hostile, extraterrestrial race may realize, why battle Humans into submission, when you can trick them into submission. They could command our limbic system, by reprogramming our brain, interjecting into us a hostile biologic program that places us into a submissive mode, possibly by activating those subroutines in our programming that stimulate reward or the qualities of charisma.

We know Human Immunodeficiency Virus (HIV) which is responsible for AIDS, infects the helper T-cells in the human body and reprograms the helper T-cells to create more virus. This reprogramming of the helper T-cells cripples the human immune system by interfering with the helper T-cells capacity to generate command signals to the remainder of the immune systems. Once the helper T-cells are neutralized, the immune system fails in its job to protect the human body from infectious agents. Without the command signals from the helper T-cells, the human body becomes susceptible to a number of infections by pathogens that normally cannot invade the body. By our current state of pharmacology, eventually the human infected with HIV dies from overwhelming infection (hopefully this will improve as our understanding of the immune system matures).

Again considering the alien visitor with an aggressive pathogen tainting his or her coattails, if the alien pathogen was in fact a virus designed to effect a change in the Human brain software, then the alien's virus might reprogram us to follow whatever commands the alien might give us. As farfetched and/or nightmarish as this sounds, in fact, it might be a very simple concept. Such a scenario begs us to use our intellect and decipher our own programming, possibly to create mental fire walls, so that Humans can remain a sanctioned, highly programmed, individualistic species in the eyes of the universe.

I don't mind sharing the universe, but I for one, would hate to be herded along like sheep, subservient to another race.

30

IMAGINATION GENERATION UNIT

The imagination generation unit (IGU) in the brain, located in the frontal lobes of the brain, is in part responsible for the unique creativity that all humans have (see Figure 52). The IGU is the sparkplug of the brain, it is the turning of fortune on the face of a Vegas slot machine when we insert a quarter in the coin slot and pull the trigger lever of the one-armed bandit. The IGU operates similar to a random number generator in a computer. One can write a software program which directs the computer to provide the program with a random number. When requested, the computer will generate such a random number to be utilized by the software program. Most people believe Vegas slot machines turn their wheels of fortune in a random pattern. Such a random pattern is generated by the slot machine's internal computer creating an array of random numbers each time the trigger lever is pulled. This array of random numbers dictates the extent to which the wheels of fortune will spin.

In the human computer, a similar function also exists. We have the capacity to generate random numbers, random combinations of colors and images. Once such a random number, or combination of color and image has been generated, the file is transferred to the scratch pad memory unit. In the scratch pad memory unit we may elect to incorporate this random number, random color and image into a project that we are working on. Each time a variation of the file is generated in the scratch pad memory, the conscious brain is queried, to ask if that is what it is requesting. If the conscious brain acknowledges it is satisfied with the result of the file, then the job of the scratch pad memory is finished. The file is moved to either the short-term memory, long-term memory or transferred into a physical form such as drawing a sketch, speaking a phrase or writing text. If the conscious brain indicates it is not satisfied with the results stored in the scratch pad memory, then the imagination generation unit will be requested to generate other random numbers, colors or images in an attempt to satisfy the conscious mind. If the creativity of the scratch pad memory still cannot satisfy the conscious mind, then the scratch pad memory may ask for help from intelligent agents, workhorses of the subconscious processing unit, described in a later chapter.

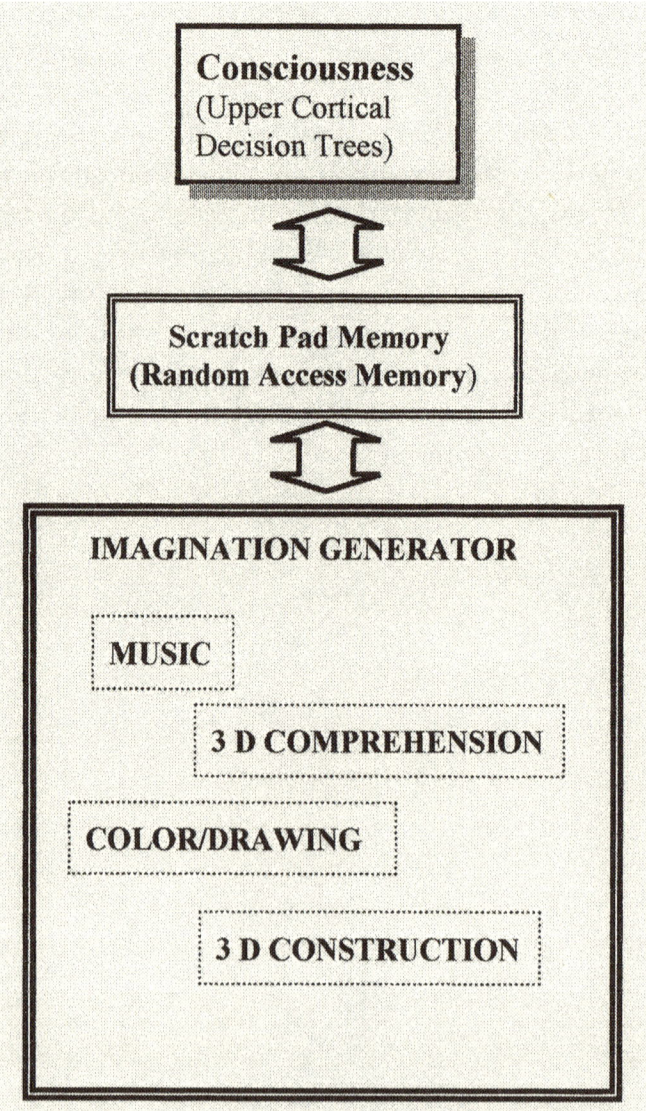

Figure 52. The imagination generation unit, the random idea generator loads ideas to the scratch pad memory for the consciousness to either accept or reject.

Years ago, I was out of town giving a lecture to the general public regarding the importance of understanding and managing osteoporosis. During the lecture I used a red laser penlight to highlight to the audience the important facts that appeared on the slides I showed. Following the lecture, I retired to my hotel room. Before retiring for the evening, I activated my cell phone and called my wife. After wishing my wife a good night, I slipped into bed and fell asleep.

At about three thirty in the morning, I suddenly woke up. I had had a dream. The dream was the vivid image of a hand-held device that combined a cell phone with a laser pointer. The imagery that I had seen in my dream was that of taking the cell phone equipped with a laser, and pointing the laser at an object such as a storefront sign. The laser beam would be modulated by the sign and return to the cell phone with the information from the sign. The information in a bar code format located on the sign would be read by an optic decoder in the cell phone. The information transferred would be the phone number of the establishment, so that the cell phone user could then dial the establishment directly, or store the number in memory to facilitate a call at a later time.

The imagery in my mind was that of the average busy person, who had two distinct tasks to accomplish before arriving home. One was to acquire dinner for the family, the other was to pick up the dry cleaning. Driving past a pizza establishment, the person in my dream took his cell phone, activated the laser and pointed the beam at the restaurant's outdoor sign. With one swipe of the sign, the phone number of the restaurant appeared on the alphanumeric readout of the cell phone. The cell phone user pressed the *send* button, and the cell phone dialed the restaurant. The cell phone user ordered a large pizza and a family size salad, while he drove off in the direction of the dry cleaners.

The imagery in my brain was so clear that I arose up out of the bed and scratched down notes regarding the idea. Being a rheumatologist (a physician that treats arthritis), my subconscious instructed my conscious brain that I might be able to generate funding to pursue research on curing the age-old disease of rheumatoid arthritis with what seemed like a new and unique idea. It was as if my subconscious brain was trying to bestow upon me a gift; a means of effecting the research I aspired to conduct. As it turned out, I later learned this concept had already been previously investigated.

The basis for literally dreaming up the above mentioned laser-phone device was obviously created by my brain loading my scratch pad memory with a string of images that had not been previously put together in my mind, and the resultant final image was nothing that I had seen before. The subconscious brain had amassed the image of the cell phone, the laser pointer, a bar code sign, and the idea of an advertisement sign that stood over the door of a restaurant that served take-out pizza.

The concept of an imagination generation unit in the human brain would suggest that at times of necessity or times of creativity, the human brain is able to take ideas, facts, figures, possibly previously unrelated to each other, sequence and combine them into a product that is new and unique and would be considered very creative.

After completing the first book of the Earth Pro series titled, EARTH PRO: *The Rings of Sol*, a fictional description of the *origin* of Earth's Master program, and then having derived the plans for *this* book, I had thought I had expended *all* of my creative energy. I felt I would not be able to write another book that would compare to the quality of the imagination expended to create the first two books. Then, one afternoon, while I drove over a set of railroad tracks on my way home, suddenly, in literally the blink of an eye, the entire idea for a fictional sequel flashed in my brain. My conscious mind saw nearly the entire storyline for a third book, base on the theme the *purpose* of Earth's Master Program. Interestingly enough, the fictional sequel neatly tied into the theme of the first two books and revealed story concepts that truly outstretched the imagination of the original story. It would seem, the imagination generation unit is capable of stringing together long sequences of ideas, to the extent of filling an entire novel.

31

GENIUS FILES—'EVERYONE HAS THEM'

I recently checked, and after loading numerous computer programs and spending many late nights writing, I still have 15 gigabytes of memory space available on my hard drive. I have been able to use up only 5 gigabytes. In addition, there are a number of programs that came with the computer, several of which I have yet to open. So despite having had my computer for several years, there is an immense area on the hard drive that can be used for file storage, and there is still a significant amount of program information that I have never explored.

Ever since I was a kid, I have heard the phrase 'humans use only ten percent of their brain capacity'. From what has been previously covered, I would estimate, contrary to popular belief, that we utilize a significant portion of our brains. I would venture to guess we use eighty to ninety percent of our brain functioning power. We take the optical signals the eyes capture and generate vision. We listen to our environment and analyze sounds with speech recognition. We smell the environment and decipher what is attractive to us versus what is repulsive or dangerous. We distinguish different tastes that our tongue receptors encounter and decide what is edible. We monitor and regulate all the internal organ functions necessary to keep us alive. We generate complex muscle movement in very precisely coordinated manners in order to speak, cough, spit or throw a ball, scratch our head, solve problems, create new art forms, new structures, new technologies, and new, interesting and often controversial forms of music.

But then, what about that last ten to twenty percent of the brain that is *not* being used? Mother nature is generally pretty purposeful. Mother nature generally does not create things in an environment that have no meaning, reason, or purpose. Why don't we have a brain the size of a peanut, like some of the gigantic dinosaurs had? Why would humans have been developed with a very large brain that would be purposeless?...Or does the part of the brain that we don't use on a daily basis hold secrets that we have yet to explore? Like the description above, does the unused portion, of the human brain allow for (1) memory files comprised of a persons own experiences to be saved and (2) store memory files that we can tap into, to advance our technology, art and science?

Inside the brain exists the conscious, the subconscious and the memory units. The subconscious interacts with the frontal lobes of the brain to work in a partnership. Our consciousness works on all of the higher level thinking that is required to make us individuals, while the subconscious toils with all of the lower level thought processes are required for the body to operate properly.

So the subconscious is the go between the consciousness brain and our long-term memory unit. The subconscious also interacts with numerous other units in the brain including the emotional unit, the sexual drive unit, the pleasure unit, the logic unit as well as the peripheral devices located outside the brain.

The memory unit is divided into four major components. These four components include Real-time Memory, Scratch Pad Memory, Remote Personal Memory, and Archived Memory (Read Only Memory).

The Archived Memory unit is quite a different set of memory files (see Figure 53 and Figure 54). This part of our memory includes data and instructions that are not of our own doing. The Archived Memory unit is broken into two subdivisions that may, on the surface, appear to be totally separate, but in actually may provide similar function to the human species. One subdivision is *Preset Decision Trees* or one could refer to this memory as Instinctual Memories, the other is *Preset Data Files*.

THE HUMAN COMPUTER

THE CORE DEVICE

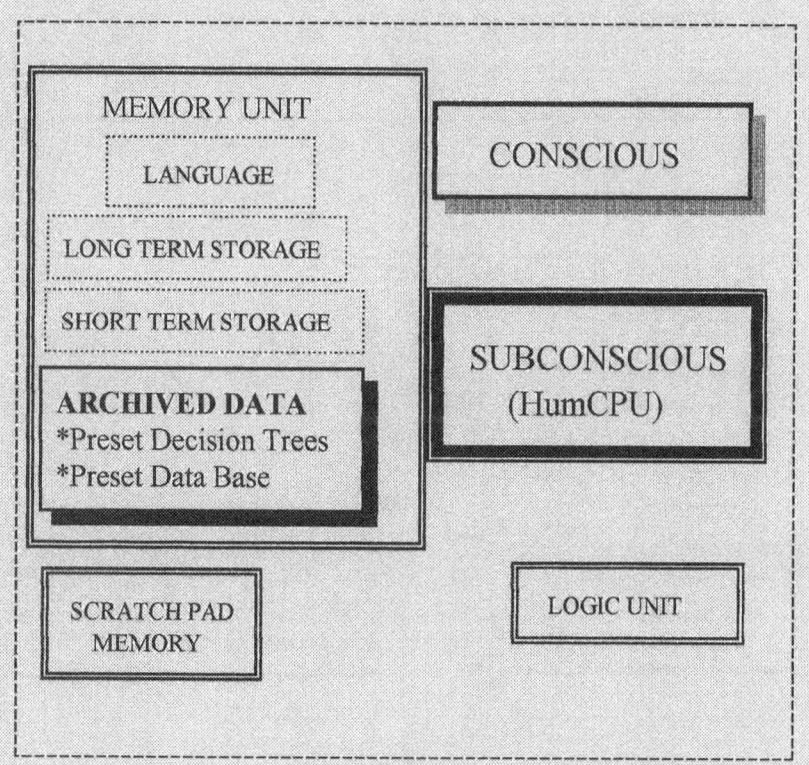

MEMORY UNIT

LANGUAGE

LONG TERM STORAGE

SHORT TERM STORAGE

ARCHIVED DATA
*Preset Decision Trees
*Preset Data Base

CONSCIOUS

SUBCONSCIOUS
(HumCPU)

SCRATCH PAD
MEMORY

LOGIC UNIT

Figure 53. Consciousness, Hum CPU, and Memory
subdivisions including the Archived Memory

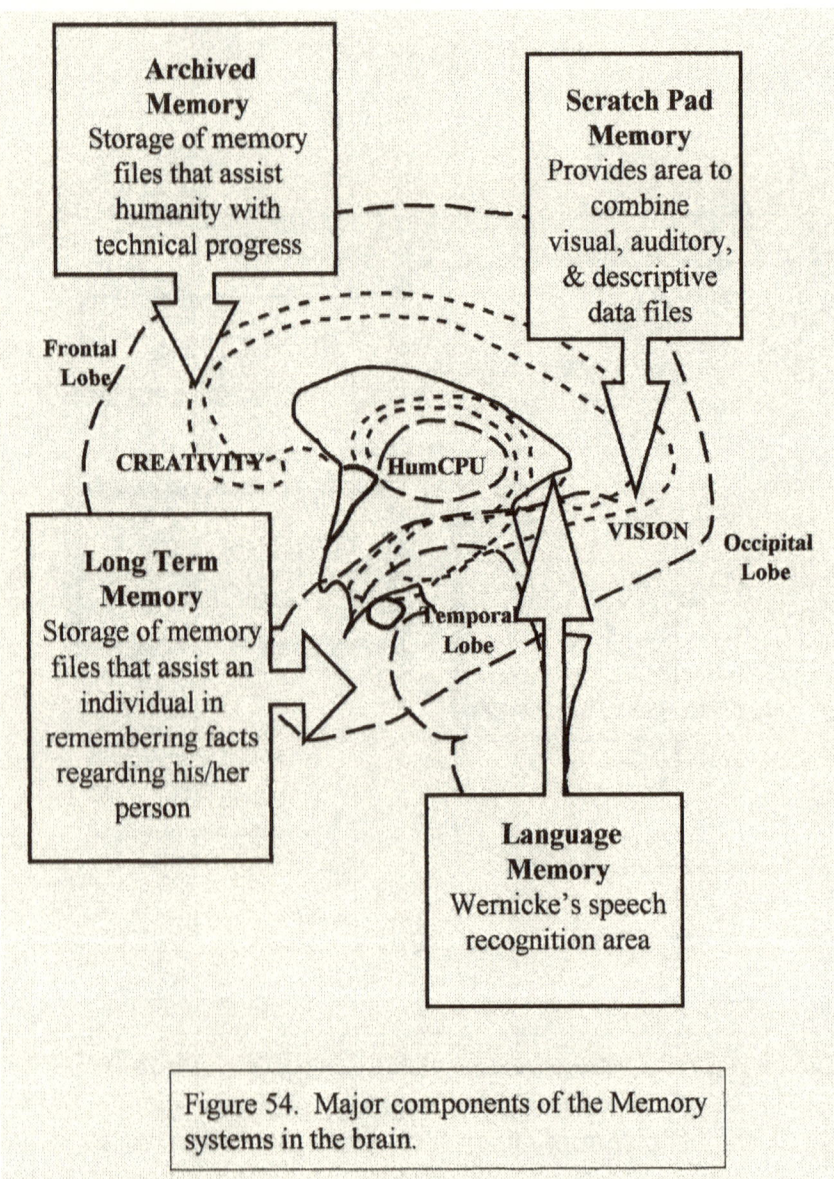

Figure 54. Major components of the Memory systems in the brain.

Preset Decision Trees are memory files of a rudimentary nature. The instinctual files are available to an individual when learned human behavior is not available or social rules and regulations are not available. These files include basic information regarding feeding oneself, avoiding danger, being aggressive to protect oneself or others, basic sexual interaction and function regarding caring for offspring. These rudimentary files are not necessarily detailed, but provide essential information for a primitive human being to survive in the wild amongst animal predators and successfully reproduce. These files would have been essential to the survival of man dating back two million years, prior to the development of civilization.

The second component of the Archived Memory could be referred to as Preset Data Files, or what we might consider to be imagination. That is, all humans contain a portion of their brain or memory, that is not actively use. Often, the right side of the brain (if you are right-handed) is considered not used. After reading the previous pages, hopefully one will realize that an entire hemisphere of the brain, does not lay dormant and that both sides of the brain (right and left) are very active. But a subsection of the brain is not readily accessible This portion of the brain may actively represent a vast library of facts and figures preset in our brains (in other words-a gold mine of new and inventive ideas, with details all the way down to the exact blueprints). This part of our brains would be referred to as the Genius Files. That *Imagination* is possibly better considered, the subconscious brain accessing a data file stored from the Genius File library in the Archived Memory portion of our brain. A good portion of the inventions mankind has devised over the centuries possibly were derived from data design files already stored in the human brain.

So the Genius Files might act like this: an aeronautics engineer is charged with redesigning current rocket engines to improve the power output of the rocket so that larger payloads can be launched into space while at the same time being more economical. The engineer sits down to figure out a solution to the task that the chief engineer has requested of him.

The engineer makes known to his consciousness the parameters of the task. The conscious mind then hands the problem over to the subconscious mind to work on the problem using facts loaded into the scratch pad memory and the logic unit. The engineer's mind sifts through the available facts loaded into the scratch pad memory. The subconscious also begins to ask the Archived Memory unit if there is any reference to the problem store away in the Genius Files. The Archived Memory searches a limited portion of the Genius Files and sends the subconscious limited information on rocket engines based on power reactors that are driven by controlled atomic fusion reactions. The subconscious loads a limited portion of the information on fusion reactors to the scratch pad memory. The subconscious alerts the conscious mind of its find. The conscious mind recognizes that controlling fusion reactions so that it can be used to power a spacecraft is well beyond the current materials technology. In short, it cannot be done. The conscious mind informs the subconscious that the idea about fusion-powered rocket engines is just not feasible. The conscious mind instructs the subconscious to work on a practical solution to the problem given the set of parameters in the scratch pad memory. The subconscious then tries to figure out a solution based on facts in the scratch pad memory. If the facts in the scratch pad memory fail to come up with a viable solution to the problem, then the subconscious will go back to the Archived Memory and request a new idea.

The response of the Archived Memory is based on the information used to request the new idea and on the basis of timing. The facts that the individual is contemplating, located in the scratch pad memory, must match up with the parameters pre-established for the new idea, in order for the new idea to be released from the Genius Memory files. Second, the timing must be right. In general, the Archived Memory looks to recognize that the level of human technologic advancement is appropriate for the new idea to be released. Once these two parameters are satisfied, in the case of the above-mentioned example, the Genius

Memory files may release to the engineer an idea related to a new rocket engine valve design that improves rocket engine efficiency, a new after burner design, a new chemical equation to be used to boost the octane of the rocket fuel, or any number of other combinations that would result in an advancement of the technology.

Therefore, problems can be solved in at least three manners. The individual working the problem can utilize (1) their scratch pad memory, experience, training and analytical skill, or (2) generate new ideas by playing with their Imagination Generation Unit, or (3) a person can have a new idea downloaded from data files pre-existing in their brain.

This scenario explains why there are geniuses in the world. Recognized geniuses may be individuals that have better or freer access to the Archived Memory files. In addition, brilliant figures in man's history such as Hippocrates, Leonardo De Vinci, Michelangelo, Copernicus, Newton, Einstein may have been individuals that trained their subconscious minds to repeatedly and actively access the Archived Memory files. These futuristic thinkers of our past may have for lack of better words, been better in touch with the Genius Files than their contemporaries.

I recall driving in heavy traffic one day headed out of town near rush hour. I was on my way to give a lecture to a group of physicians regarding the advances in the treatment of osteoarthritis. I had planned enough time to make my trip, but I was two hours from my destination and my car was stopped in traffic. As my vehicle inched along in the line of traffic, my mind drifted to conjuring up a better way to travel. I thought a helicopter would be nice, but not unique. My mind then conjured up the image of a car powered by a fusion reactor for an engine.

I thought this would be great, but technically extremely difficult to accomplish, difficult to control the rate of the fusion reaction, and difficult to create enough insulation to shield the human passenger from the resultant radiation.

As the traffic crawled forward, step-by-step, my mind tried to define what it was I was attempting to overcome. Gravity was holding my vehicle's wheels to the surface of the road. What I really wanted to do was defeat gravity. I wanted my vehicle to lift up, above the other cars that stood in line before me, and sail off to my destination.

My analytical mind drifted off to ponder what was the essence of gravity. Not being a physicist, in my own limited universe, for all I knew, no one had yet truly arrived at a good, concrete explanation for what caused or generated gravity. Yes, I was aware of the theory that molten metal supposedly continuously churns in the core of the planet, which possibly like a current passing through a coiled wire, sets up a magnetic field, that acts as gravity; But I had always wanted a more complete explanation.

As I pondered the idea of gravity, my scratch pad memory became active. I had recently completed a science project with my middle child. This project had involved a solar panel. We had generated current in a circuit using various light sources. In our circuit we had an electric motor. The blade of the motor spun when a sufficient current ran through our solar circuit. Explaining how the motor worked to my son, we had discussed that when an electric current passed through a coiled wire, this created a magnetic field that exhibited an attractive or repulsive qualities depending upon the direction the current flowed.

My scratch pad memory took an enormous leap of faith, as it considered that the material of the Earth's surface is made up of atoms comprised of electrons that all spin in one direction. At the center (core) of the planet, may exist a powerful material that is comprised of atoms that are comprised of electrons that spin in the opposite direction. Now it is true that an electron revolves around the center core of an atom, and does this with such speed it produces a cloud. The more protons at the center of an atom, the more electrons comprise the cloud of the atom (protons are considered positively charged, electrons negatively charged). Still, the direction the electron is traveling, in general, creating the cloud, may be the determining factor between what is *matter* and what could be considered *anti-matter*.

Therefore, the atoms comprising the bulk and the surface of the planet are attracted to the core of the planet by the fact that the atoms at the center of the planet are comprised of electrons that spin in one direction and the atoms on the planet's surface comprised of electrons that spin in the opposite direction. The opposite, and therefore attractive force created by the two different types of material, exhibit an attractive force upon each other. This attractive force between two materials with opposite spins is what is known as gravity (closely related to magnetism-created when current runs through a coiled wire), and in essence holds the atoms comprising the planet together (see Figures 55). If humanity could take matter on the surface of the planet and effect a change in the orbital rotation of the electrons of an atom, then the attractive force of gravity may be transformed into a repulsive force-Wow, my imagination was really in overdrive on this particular day!

Consider then, free energy could be created by simply changing the direction of a material's electron spin, then changing the electron spin back causing the material first to repulse gravity, then to be attracted by this same gravity, which could constitute a piston function engine where energy is generated without having to burn or otherwise consume fuel as required in a conventional engine (see Figure 56).

The final result could be to produce a vehicle whose engine produces anti-matter that would defeat gravity by being repulsed by it, causing the vehicle to lift up off the surface of the planet (see Figure 57). This technology could lead to a fuel-less, radiation-less, heat-less power supply for interstellar travel. Now the above is purely science fiction, none of which is anchored in fact. But just thinking about the possibility, might lead to ideas that are substantial, and could lead to a solution to our present need to burn fossil fuel and at some later date, our need for interstellar travel.

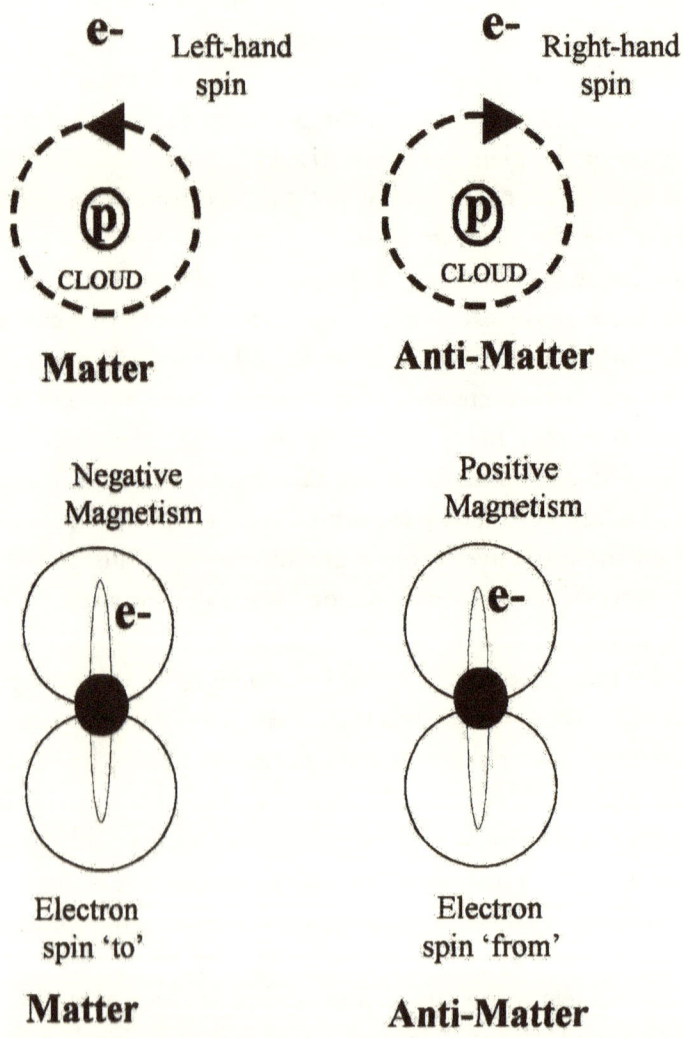

Figure 55. The state of Matter versus Anti-Matter depends on which direction the electrons spin around the nucleus of the atom.

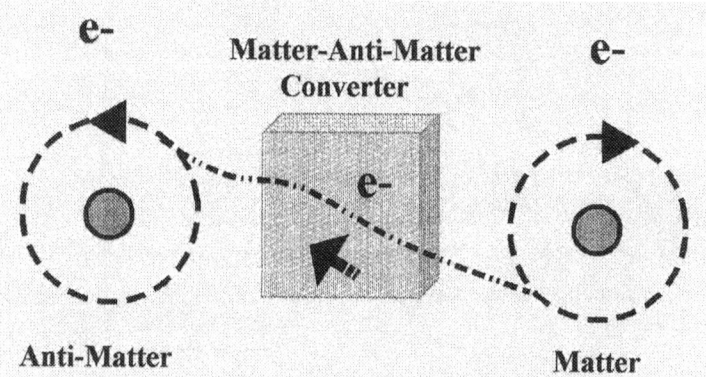

Matter-Anti-Matter Converter

e-   e-

e-

Anti-Matter   Matter

Device Changes electron spin from left handed
(Matter) to right handed spin (Anti-Matter)

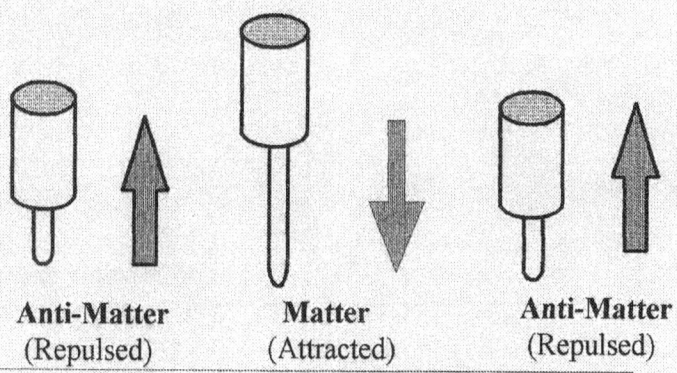

Anti-Matter   **Matter**   **Anti-Matter**
(Repulsed)   (Attracted)   (Repulsed)

EARTH SURFACE: Matter

EARTH CORE: Anti-Matter

Figure 56: A <u>free energy piston engine</u> could be made by
converting matter to anti-matter and then back to matter; and
taking advantage of the Earth's magnetic field.

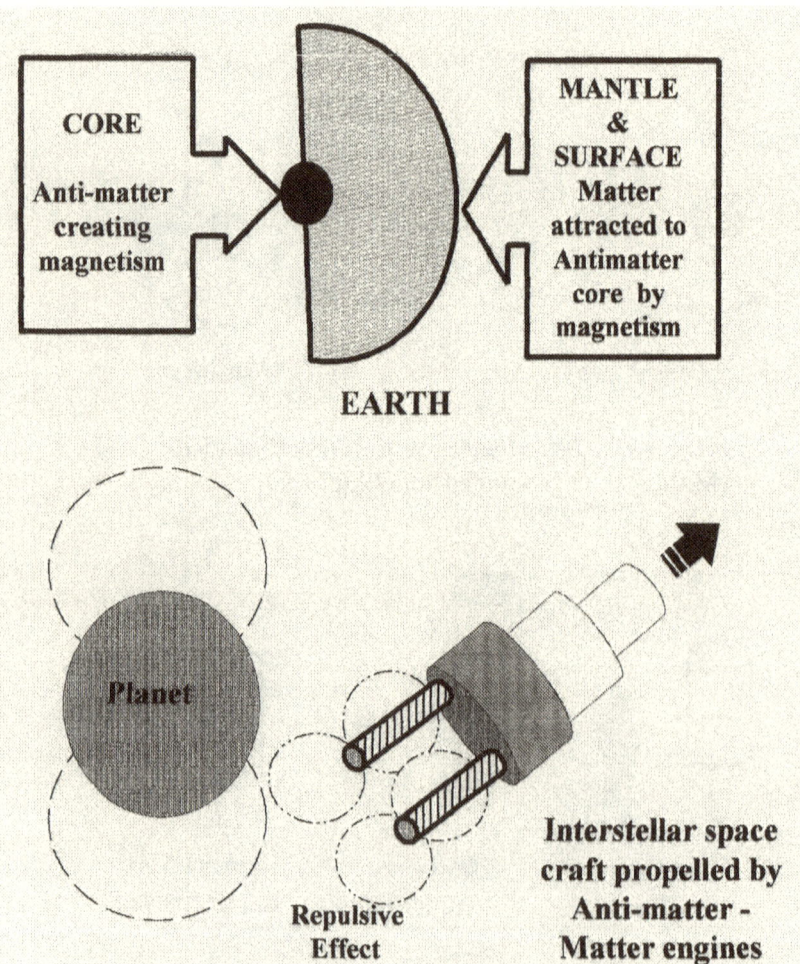

Figure 57. Anti-matter creates planets by attracting matter. Anti-matter engines could be used for space travel: by causing a repelling effect to push ships away from planets, and later changed to matter engines to create an attractive effect to pull space ships toward distant planets.

What is curious is that the imagery described above, flooded, in its entirety, into my conscious mind while I sat, my hands clutching the steering wheel, my car inching along on pavement behind a hundred other vehicles. Possibly, my brain did encounter too many exhaust fumes from the cars and eighteen-wheelers packing the highway. But still, reaching any explanation that described a possible fundamental principle for a natural phenomena like gravity, while stuck in what felt like a deluge of traffic, was, well satisfying at the time,...better than being frustrated that I wasn't going too far to fast.

Genius files may be difficult to accept as a concept. The human being thought of as a computer may also be difficult or even impossible to believe or accept.

In the first two years of medical school part of the study material included embryology. Embryology is the study of how a single sperm cell from a man fuses with an egg cell from a woman, and develops into a human fetus. During the many stages of development from the two sex cells to a complete human baby, the human fetus undergoes drastic changes in its appearance while in its mother's womb. The original fused cell referred to as a 'zygote', divides into immature cells. The subsequent cells divide over and over again. Distinct cell lines including the brain and nervous system, heart, gut, muscle and bone tissues, develop from the initial core of immature cells.

The fascinating fact is that this intricate, nine month, highly sophisticated process, has been repeated billions upon billions of times over a span of hundreds of thousands of years, much of the time providing the world with a similar product...a normal healthy human baby. The fact that this process has been repeated so many times over so many years with such precision, suggests that the process of embryology is, in fact, a biologic computer program.

Being an electrical engineer, the study of embryology fascinated me. I had written numerous computer programs as a high school student, an undergrad and as a staff electrical engineer. The fetal development process I saw unfold while studying embryology, clearly appeared to me to be guided by a biologic software program.

Modern science had photographed or simulated in artistic representations, each step of the process of embryology, from the point of two sex cells fusing into a zygote to what we recognize, nine months later, as a human baby. As a student, new to medicine, armed with the creative energies of an engineer, my mind played with the idea of embryology. A fascinating fact was that the success of the process from single celled zygote to human fetus was strictly dependent upon an uninterrupted cell division. Fraternal twins are twins that develop in a woman's womb from the fertilization of two separate egg cells that develop in the uterus at the same time. Amongst fraternal twins, the sex of each twin may be different. Identical twins on the other hand are identical in appearance and sex.

Identical twins share the same zygote cell, but as cellular division occurs, early in the division process, somewhere a separation occurs and two or more distinct cell groups form which eventually results in two or more identical twins. What is fascinating is, when a separation of the initial cell group occurs, the system doesn't end up as a failed attempt, nor does it produce an end result that is deficient in some manner…That is, identical twins are not generally recognized as being inferior in intelligence or physique when compared to other humans. Therefore, there must exist some supervisory or feedback mechanism in the early cell group that causes structures to develop that might otherwise be lost when the original core of cells separates into twins.

As mentioned above, the early cell group develops into a nervous system, a heart, gut, bone and muscle tissue and many other specialized tissues. Along the path of development, cells differentiate. In the nervous system, the cells eventually differentiate into the brain cells, the spinal cord and all of the peripheral nerves and special senses. Finally, some mechanism must exist to tell the cells when to stop differentiating. A head should only be so big. An arm and a leg should only be so long. The entire size of the average fetus can be only so big, otherwise travel out the womb through the birth canal could not occur. Some mechanism causes ears to appear on our head and not on our butt, hands to occur on our arms and not on our legs, a nose to occur on the front of our face and not on the back of our head.

Being an electrical engineer, my subconscious often toyed with trying to identify some rational explanation for how the developing embryo successfully developed into a fetus destined to be born as a baby that we could all love and cherish, because the baby has an appearance similar to ours. It is a very unique puzzle, given the level of sophistication of the process.

As a third year medical student on a surgical rotation, you started your days waking at four in the morning. You arrived in the hospital at four-thirty in the morning so that you could visit all of the patients you had been assigned to monitor. At six in the morning you rounded with the surgical team, consisting of a senior surgeon, a senior surgical resident, two interns (first year residents) and two or three medical students. The team interviewed every patient that comprised their service. Rounds usually ended by seven. The surgery for the day usually started at seven-thirty. Surgical cases were conducted all day. The team usually reconvened at six or seven o'clock in the evening to review the condition of all the surgical cases that were on the team's service. Often, all of the patient's were visited. Rounds often lasted till eight or nine o'clock every night. Once rounds were done, those that were not on-call could go home as long as their work had been completed. Those assigned to be on-call stayed the night in the hospital. The on-call members of the group attended to the medical and surgical needs of the team's patients as well as other surgical teams' patients. Generally, on-call assignments were on a four day rotation. Every night, one team would cover all night for the other three teams.

I recall standing on the seventh floor of the hospital. It was my night to be on-call with one of the surgical interns. I was dressed in the pale green scrubs, which I had worn all day. It was ten o'clock at night. The other three surgical teams had left for home. The intern I was on call with headed for the cafeteria in search of a snack. I stepped into the cramped quarters of a back room, that acted as the resident's room, where we left our personal items during the day. I remember picking up my rucksack and turning toward the door. Suddenly, it seemed as if a door opened up inside my head.

For only a brief moment, it seemed the entire theory behind the physiologic mechanisms of embryology came rushing into what I would consider my conscious brain. I remember standing for a moment, in complete awe. I was suddenly reviewing, what seemed to be the very blueprints of how a single fertilized egg developed into a complete human infant. It was as if for two years, I had asked my mind the question *How does embryology work?* and I had irritated the subconscious enough that my brain had searched the distant recesses of my brain's memory for the answer to this recurring question. On this particular night, at that particular moment, my subconscious decided to deliver the answer to my nagging question, only it didn't just deliver a half-baked notion on how embryology worked, it seemed to have delivered the blueprints responsible for program logic of embryology, in a detail well beyond what I had seen or studied to that point in my medical education. For what seemed like a fleeting second, I felt I knew everything there was to know about embryology. Then as fast as it came, the vivid imagery left, the open door shut…the feeling or concepts faded, never to surface again.

I felt like Bones, the Enterprise's physician, how he felt in the Star Trek episode where Bones interacted with an alien intelligence, and briefly gained the knowledge to surgically reinsert Spoc's stolen brain back into his head.

I was so empowered by the remnants of knowledge left in my conscious brain, I took the time out of my hectic medical school schedule and wrote a paper regarding the facets of embryology that had been revealed to me. At that time, I had access to only a typewriter, not a word processor. I compiled my typed essay and the drawings into a thesis. I presented the work to an anatomy teacher at the medical school. The work probably never got beyond a casual review, and eventually lost when the school decided to relocate the anatomy professor's office to a different part of the research building. I'm sure the paper met its end in a circular file. I must admit though, the idea was very much off the wall. Could you imagine, eighteen years ago, stating to anyone, that a human infant was the result of a biologic computer program?…What a radical idea, even for this new millennium!

To this day, I distinctly recall how I felt when that door opened up in my mind. I recall how empowered I felt witnessing what seemed like vast volumes of knowledge. I unfortunately do not remember what it was I saw with my conscious mind that night. The images deteriorated rapidly following the experience, and have faded over the years. I do recall the sense of disappointment and the incredible sense of loss I felt, when the door shut and I was left with only my own mortal thoughts. I have never had a similar experience…But as far as believing in genius files…Yes, I do,…And I believe that every individual has them. We need to recognize that genius files exist, and in that instant we may be privileged to the knowledge base present in the deep recesses of our minds, pay attention to the knowledge and let it empower you to accomplish great things for yourself and for humanity…Oh, what secrets and treasures the depths of the human mind might hold!

32

INTELLIGENT AGENTS

Crawler programs are what an Internet search engines use to develop data files. Search engines utilize crawler software to explore the Internet. These creepy crawler programs scour the Internet for facts about the Internet. The creeper programs then relay back to the search engine's program database where information is located out in the Internet regarding subjects that search engine is interested in storing and offering to its customers.

When a customer of a search engine requests to be directed to a web page on the Internet regarding a particular topic, the search engine scans the database it has in memory. What is displayed to the customer on their computer screen is the list of web pages that the creeper programs have found in their searches. The list is ranked in order by whatever criteria the search engine uses to decide which web pages are listed first versus last.

The customer scans the list of web pages on their screen and when the customer decides that they would like to explore a particular web page and clicks the mouse symbol on their choice, the search engine then points them in the direction of the Internet address of the web page.

Intelligent agents are the *intelligent* next generation of data analysis computer programs bounding way beyond the concept and capabilities of the crawler program.

To understand what an intelligent agent is, we first have to explore what is the meaning of an agent, and what is the meaning of intelligence as used in the computer world. The term 'agent' may refer to a software subroutine or small program that carries out one or more functions on behalf of a person, another program, etc. The word 'intelligence', as described by the dictionary, refers to the capacity to acquire and apply knowledge. Intelligent agents, an offshoot of artificial intelligence, brings an entirely new concept to the world of computers, and possibly explains a vital processing function that operates in our brains.

The world is exploding with megatons of data as the information age rolls into high gear. As each consecutive moment ticks off the clock in real time, vast amounts of data are being generated on a global scale relating to all facets of life. The gathering, organizing and storing of large amounts of data can be referred to as *data warehousing*. When a human decides to analyze the information and extract interesting pieces of information from the database this can be termed *data mining*.

So for example, the owner of a hardware store may have his inventory computerized. When he puts his overstocked inventory of hammers on sale he may extrapolate following the sale, information regarding other items customers purchased when the hammers were on sale. He might find that large numbers of nails, wood products, saws, and sand paper were sold when the hammers were put on sale.

The above would be an example of data warehousing in which the data from the date of the sale of the hammers was collected and placed in a computer database. An example of data mining would be the storeowner physically searching the database for items that were sold in larger than normal quantities on the day of the sale of hammers. After repeating this exercise over several sales, the storeowner might come to the conclusion that if he places his hammers on sale, to increase profits, he should have a sufficient supply of nails, wood, saws, and sand paper in his inventory ready to be sold to his customers because review of the database suggests these are what the customer base will buy in addition to the hammers. The storeowner is exhibiting 'intelligence' by demonstrating that he can search the database and apply the knowledge he has gained from the search of this database to his sales strategy to make his business more successful.

The problem humans face is, we are not well equipped either in 'interest' or 'brain processing function' to routinely decipher large databases and make useful judgments from these analyses in a timely fashion. Staring at vast amounts of seemingly unrelated data and trying to determine trends in real-time is a daunting task for most of us.

The term *intelligent agents* refers to a software concept currently emerging in the computer world. Where artificial intelligence generally refers to an extensive software program or set of software programs written to act as a global intelligence capable of commanding all of the functions required to operate a complex computer system such as a mechanical robot device, the concept of an *intelligent agent* refers to a software program written to perform possibly a single intelligent function and perform the function quickly, as in real-time.

Returning to the example given above, instead of the storeowner himself searching the reams of information in his data warehouse, he might employ a specially written computer program that would search the database and return the same conclusion the storeowner had come to; the conclusion being that when he places his inventory of hammers on sale that he also sells excess amount of nails, wood, saws and sand paper. Such an intelligent agent software, not intimidated by the vast amounts of data it may have to search, may, in addition, reveal that other items such as excess amounts of masking tape and drop cloths were sold a week later. A sophisticated intelligent agent may reveal that two weeks after the hammers were put on sale an excess amount of paint and paint brushes were sold in the store.

By using an intelligent agent to analyze the database, the storeowner may obtain a more reliable picture of his customer's buying habits. The above example may really be demonstrating that when the hammers are placed on sale, the customer base is encouraged to purchase the hammers, and in addition purchases nails, wood products, saws and sand paper to initiate a household project. The purchase of masking tape and drop cloths a week later and the purchase of paint and paint brushes two weeks later are a reflection of construction projects being completed by the customers and now the customers are exercising their desire to put the finishing touches on the construction projects by painting them. Now certainly not all customers would be capable of starting and completing their project in the suggested time frame, but the intelligent agent would be able to identify the purchasing habits of the average consumer.

Such intelligent agents could also alert the storeowner to defective equipment. If a week after the hammers were placed on sale, a number were returned because they were broken, then the storeowner may elect not to offer that brand of hammer to his customer base in the future.

But how do the concepts of data warehousing, data mining and intelligent agents relate to the human brain, except for the fact that we have identified above that the human brain is not terribly eager or efficient in analyzing large databases in real-time. This deduction of the human processing capacity seems correct when referring to the conscious processing power, but may be quite incorrect if applied to the subconscious processing powers.

The subconscious brain may engage numerous intelligent agent-like biologic software programs to assist in the tasks the conscious brain asks of the subconscious brain. Problem solving persons such as engineers, are well acquainted with the process of gathering facts regarding a problem and tasking the subconscious brain to render a solution to the problem. A problem-solving individual may have the parameters of such a problem sit in the subconscious brain for minutes, hours, days, months or even years before the subconscious brain processor delivers to the conscious brain a viable solution. Such a solution to a problem may appear in the conscious brain at any time, including waking the person up in the middle of the night if the problem to be solved is of an urgent nature.

An example of such an occurrence might be that of an engineer assigned the task of building a dam across a wide steep walled canyon with a sizeable, rapidly moving river passing through. The dam is required to produce electrical power to the surrounding communities and prevent life-threatening floods caused by runoff water from mountains up stream during rainy seasons. The problem is that though the available technology can accommodate concrete drying under water that is still or slowly moving, the intense pressures caused by the force of fast moving canyon river would prevent concrete from ever drying properly. The civil engineer assigned to build this dam knows that he needs to temporarily divert the river, but he is unable to technically divert the water flow of the canyon river to another waterway because the river starts as an underground spring with no tributaries upstream. The canyon walls are thick, steep and solidly built. The canyon walls are too thick to allow the water to escape the canyon. Though others have claimed that the project is impossible, the civil engineer's mind analyzes the parameters of the problem and eventually arrives at a solution that involves blasting a diverting tunnel through the steep walls of the canyon to bypass the site where the dam is to be built. Special blasting techniques are derived, special mining trucks are fashioned to work in the steep ravine. After organizing his data, the engineer presents the solution to his superiors. The plan is approved, the blasting is done, the diverting tunnels are cut through the canyon wall, and eventually the dam is successfully built. In this example, the civil engineer is responsible for arriving at a means of diverting the canyon river so he can effect the construction of the dam. His brain's intelligent agents analyzed the data and arrived upon a solution that though technically difficult and time consuming, did allow the dam construction to proceed forward and eventually led to the successful completion of the dam.

Solutions to difficult or complex problems may be performed in the human subconscious brain by intelligent agent programs. Intelligent agents may be loaded with data parameters and then scour a person's memory files and interact with a person's logic unit to arrive at a solution to the assigned problem. The entire concept of being intelligent, that is being capable of applying knowledge, may be related to how efficient or efficiently trained the intelligent agents in our subconscious brain operate.

Looking back, my impression of my undergraduate experience was that much of it was way over my head. Thinking back on it though, the school I attended was very concept oriented. At the end of my electrical engineering training, I had no idea how to fix my television set if it ever broke, but I did know how to problem solve. The Electrical Engineering department's objective was not to produce engineers that could put together things other people had designed, but to produce graduates that could design things that had never been conceived of before.

I guess in reality, the curriculum I endured, helped me learn how to train and utilize my 'subconscious', my mathematical processors, my scratch pad memory and my 'intelligent agents' to problem solve, so that I could survive in the real world. That made school well worthwhile.

33

THE ALL IMPORTANT SECRET TO PROGRESS: 'THE VARIABLE CONSTANT'

No, this is *not* an oxymoron, but *instead*, a *vital key* to human survival.

Music is an ever changing part of a free and uninhibited society. With each up-and-coming generation, the music changes. The bible embraces music as a means of human expression. It turns out that music may be a very important influence for the developing brain.

As the prefrontal cortex of the brain (the front portion) matures, the decision trees, operating pathways, and learning circuits in the brain undergo intense development and refinement. Imagination and problem solving capabilities define themselves as the prefrontal cortex ages. Music provides a defined data stream that can be repeated over and over again, and this offers a stimulus for the developing human brain to work on to help refine certain functions of the brain. The sounds comprised of the melody and the lyrics act as a repeatable, variable data input that subconscious processor, the learning circuit, the septal region (pleasure center), and the prefrontal cortex can decipher, analyze, record, and playback at any time. The lyrics linked to the melody may not have to maintain a theme or make sense to the casual listener. On the other hand, the artistic flare of the songwriter may stimulate moods and emotions, and conjure up visual images in the form of accessing previous memories that stimulate an individual's thought processes. The memory files that are linked to the music's melody and/ or the verses of the song may trigger the pleasure centers of the brain when the brain entertains itself by listening to music.

Music though, may have a much more important role for the individual and society in general, than just simply to stimulate our learning centers or create a form of pleasurable entertainment.

Computers rely on an internal clock to set the timing on how they run. This internal clock evenly paces out the functions for the computer. With each click off the internal clock, the computer knows to progress its functions ahead, or in terms of the computer software, continue executing the lines of computer code.

The suprachiasmatic nucleus (SCN) acts as the biologic clock in the human brain. Pulses from the SCN guide the human over a predictable 24 hour cycle, the circadian rhythm. Like a computer's internal clock, the SCN paces out the functions of the human body. But what about the grand scale of human progress?

In the human computer, the advancement of technology also requires some form of a clock to execute new ideas in an orderly fashion, otherwise pandemonium would occur. Humans often seem threatened by change. The introduction of new ideas that cause change to existing paradigms are not necessarily greeted with acceptance, but more often with resistance unless the change to this technology is drastically needed, such as in wartime.

To bring about an orderly advancement in human technology requires taking the pulse of the current technology, and interjecting new ideas at a rate that does not overwhelm the existing technology and knowledge base of the population. To accomplish this vital task, the human brain needs a form of a clock. Such a clock based on time, similar to that used by conventional computers, would be inefficient.

The advancement of human technology, once set in motion, could not be effectively tied to a time clock because the rate at which humans would progress their technology would be unpredictable in real-time. Natural disasters, the ravages of war, the oppression of governing bodies affect the speed at which humans are able to advance. A drastic blow to human progress occurred when the great Library at Alexandria was destroyed by barbarians. A complete slow down of human creativity in the Europe occurred over hundreds of years with the daunting oppression created by the Dark Ages. Much of our opportunity for advancement has been forever lost when racial purging occurred during fierce aggressive periods of human history.

So, the advancement of human technology requires something other than a time clock in order to progress in an orderly fashion. Since the state of the capacity by which humans will be able to sufficiently accept the introduction of a new idea, and have the technologic capacity to utilize and incorporate a new idea is unpredictable, there needs to be a constant that varies with the technologic advancement of society. If the technologic advancement is slow, the variation in the constant is slowed. If the technologic advancement rapidly increases, then the variation in the constant also increases and the human brain is capable of sensing the variation in the constant and appropriately advance to the next step of ideas.

The challenge is to determine what in the environment could function as a *variable constant*. The variable constant would have to be something that could flow unrestricted throughout the population. It would have to be something that was not language dependent. It must be capable of being preserved through the ravages of war and natural disaster. It must be something that humans can carry with them and something that is acceptable to society and even possibly integrated into social functions to increase the dissemination of the variable constant signal to as many human brains as possible. The variable constant must be nonthreatening, even pleasurable so that the society will accept its presence, and accept the changes that occur in it while it is present within the fabric of society. It would need to be something that was tied to technologic advancement, yet not necessarily perceived as a form of technology.

The ultimate puzzle…What indeed acts as the clock humans vitally need?

Answer: such a **variable constant** is '**Music**'.

Music advances with society. It disseminates through the fabric of society. It changes as we advance technologically because the materials and design of the instruments change as technology advances. It is not that old songs or music are less pleasurable or stimulating. New songs stir different emotions and bring about an advancement in our creativity. If anything, old songs lock us into the emotions and memories that occurred during the particular time we first heard the piece of music. As each generation passes, they associate with a particular form of music. The music is heard and imprints on the person's mind events that occur with the music.

Music also offers the computer in the human brain a clock, that tells our subconscious processor when it is appropriate to release new information files to the creative members of society, so that society can advance its technologic base. The interpretation of music is located in the temporal portion of the brain near where long-term memory is stored.

Music offers a sense of pleasure and encouragement by stimulating the brain waves of the population. This is why music has been long recognized as an important art form by most societies. The upper class citizens of society often go to great expense to fund music as an art form. In modern American society many forms of music have flourished, and in general, music has been a very successful and lucrative industry while enjoying a widespread impact on society.

Most religions embrace a Doomsday scenario for the end of the world. Despite the fact the Earth has been in existence for 4.6 billion years, humans at best, have lived on the planet for only the last two million years. If the Doomsday scenario is one of a judgment, whereby we are accepted into heaven judged by our good deeds versus evil nature, there is no need to have industrialization, technology or progress. Humanity should remain farmers, sheep herders and nomads with only a primitive government to adjudicate human behavior. A lack of technological advancement would have cut down on one of humanity's greatest enemies, environmental pollution, and on the perpetration of many of the supposed sins that seem to befall us.

But if the Doomsday scenario is in fact a puzzle hidden inside this very common prophecy posed to humanity,…A puzzle that we need to solve, because species survival ultimately depends upon solving the puzzle in a timely manner, then technology maybe the key. Humanity may in fact be in a race to defend or to escape this planet. A physical time clock may be ticking away, either driven by overpopulation of the planet, pollution, some other disaster buried inside of the planet, or some celestial disaster headed on a collision course with our planet. Technology and exploring the secrets locked away in the yet unexplored memory spaces of our brains, may our best chance for survival. Hopefully, the clock run by Music, is running faster than the time-line of the Doomsday prophecy.

34

INSTINCT: STEPPING STONE TO MOTIVATION

Most animals live from moment to moment. Animals exist in a life and death struggle from the moment they are born to the second they die. Many offspring of a species die early in life, falling prey to larger animals—thus supporting the food chain. Since most animals have an uncertain future, there is in general no reason for an animal to possess or harbor 'items' while they are alive, except for a meal they have killed.

On the other hand, most of us are aware that squirrels exhibit a pattern of behavior that includes the gathering nuts for the winter. Squirrels collect nuts and burying these nuts in the ground. Squirrels do this so that the nuts can be retrieved during the winter months when food is scarce. Squirrels face death every moment of their life, but yet they exhibit this behavior of storing nuts. The behavior of storing nuts by squirrels is recognized as a simple act by casual observation. But the act of gathering and storing nuts represents a very complex behavior pattern, in-the-fact that it represents *delayed gratification*. Delaying gratification to an animal, in the wild, where survival for the animal is a minute-to-minute venture, suggests a strategy has been programmed into the animal.

When we consider that animals possess 'instinct', the concept of 'instinct' represents two different behavior patterns. Animal *instinct* on one hand is considered *how an animal knows to perform tasks that account for the survival of themselves and their species*, and *how an animal*

289

might assess a situation to predict an outcome so as to choose the best course of action to improve their chances of survival.

The behavior of a squirrel to gather nuts and bury them is considered, by most, to be *instinctual behavior*. Squirrels gather nuts and bury these nuts because 'instinct' instructs the squirrels to perform this task. Recognizing that instinct is responsible for the squirrel's behavior suggests a biologic program inserted into a squirrel's brain dictates the behavior.

Recognizing that a biologic program exists in squirrels that dictates a behavior that represents *delayed gratification* suggests that primitive animal life possess the early biologic programming necessary for the more advanced behavior pattern that humans exhibit known as *motivation.*

Humans exhibit behavior patterns based on delayed gratification every day of their working lives. We toil at our jobs hoping to get a paycheck either when a particular task is completed or at some set interval as prearranged with an employer. Many of us save some portion of our earnings for our retirement in IRAs or other funds. Many of us tend to save money over a period of time in order to effect the purchase of an item significant to us such as a car, a house, a boat, a new coat or other item we may value.

Motivation is the pattern of behavior whereby a human sets a goal, defines a strategy to achieve the goal, then works towards obtaining the goal. Motivation therefore, represents a complex, multistage behavior pattern for humans.

To consider the steps needed to produce motivation one must consider that (1) a goal must be defined; (2) the object of the goal must be loaded into the target acquisition register as an object the body is seeking to acquire; (3) strategy, comprised of steps the body must complete in order to obtain the goal, must be defined and set into a memory file which can be referred to on an as needed basis; (4) the pleasure center of the brain must be adjusted so as to respond by releasing endorphins (substances that act as triggers to stimulate opiate (narcotic) nerve receptors) to provide pleasure to the body when the goal or portions of the goal are achieved.

With regards to the squirrel, the program steps that comprise the biologic program would include recognition of the time of the year. When Fall occurs, the squirrel's target acquisition register would be loaded with a file that represents the image, smell and/or taste of a 'nut'. The squirrel would search for nuts on the ground. Once a nut is located, the biologic program would transfer to a subroutine that would instruct the squirrel to bury the nut. Once the operation is completed, the pleasure center in the squirrel would release endorphins to cause the squirrel feel good about his behavior, thus reinforcing the behavior and motivating the squirrel to locate additional nuts.

With regards to humans, the 'goal' or target of our motivation may be concrete such as earning money. The goal may be more indirect such as earning money in order to buy a new record album put out by a favorite musical recording artist. The goal may be more abstract such as working on good grades in high school in order to gain acceptance into the college we would prefer to attend. Certainly, when we receive the letter in the mail that congratulates us on being accepted to the vocational program or college or graduate school or other program we have an interest in, we feel pretty good about ourselves as the internal endorphins generated by our brain race through the vast network of blood vessels in our body.

35

THE ARTIST AND THE ENGINEER

The brain of an individual is set up with a consciousness that is able to know that the body is alive, able to receive data inputs from various sensors, and able to process information to make decisions on how to act. Then if the consciousness chooses, it is able to act either by manipulating the motor muscles, in the limbs attached to the body, or express itself through facial expressions, voice synthesis, writing, typing, waving or manipulating an instrument.

The average brain accumulates sensor data through the eyes, ears, nose, mouth and the peripheral position, touch, temperature, vibration, and pain sensors. The majority of the routine information is filtered by sensor nuclei located both in the brainstem and the subconscious brain processor. Only important data is routed through the subconscious to the consciousness, such as someone calling the person's name, a sudden 'bang', the wailing of a siren, ringing of a phone, pain generated by a thorn pricking a finger, or a flashing red light. Important information is scrutinized by the consciousness. Depending upon the interests or agenda of the consciousness, the person may wish to ignore the incoming data or may choose to further investigate the data.

A person from behind you calls your name as you are leaving your house. The ears detect the sound along with all the other ambient sound that surrounds the individual. The incoming sound signals are transferred to the subconscious. The subconscious filters out the ambient noises of a toilet flushing, a sister sneezing, the wind lightly howling outside. The subconscious separates the voice pattern using a frequency filter that detects the harmonic frequencies generated by voices. Having recognized the sound pattern of a voice, the subconscious works on three processes simultaneously. First, the meaning of the voice pattern in terms of a text sentence, second the origin of the sound pattern both in the context of who could be sending the sound signal and from where the sound signal could be being sent, and third the volume or urgency with which the voice signal was generated. The subconscious recognizes that the name being called is that of the person's sister, the sound is generated from someplace upstairs in the house by the person's mother, and that the tone is that of casual conversation. The subconscious filters out the sound and does not alert the conscious mind of the incoming voice signal, except to register the sound as an event in the person's conscious memory.

A second later, as the person is stretching out their arm to reach for the knob on the front door, the ears detect another voice signal. This time the subconscious recognizes the voice pattern its own name. The subconscious recognizes the voice pattern to have been generated by the individual's mother, the origin of the voice pattern is from upstairs in the house, the exact location unclear. The subconsciousness detects a high volume and a sense of urgency in the tone of the voice pattern. The subconscious transfers the message along with the analysis to the individual's consciousness. The consciousness elects to halt forward motion of the body's legs and halts the attempt by the right upper extremity to reach the doorknob and turn it. The consciousness reassigns the locomotion of the body to turning around, generate a verbal response and replying to the mother's inquiry.

In the average brain, information is gathered by the body's sensors and transferred to the subconsciousness. Important information is relayed to the conscious portion of the brain. Information that needs to be remembered, such as a new acquaintance's name, phone number or address, is initially stored in short term memory, later transferred to long term memory if the data is important enough to be remembered. As a person attends to tasks, such as cooking, cleaning, or working at a person's job, memory files are constantly accessed for stored information that tells the person how to complete the tasks the person is attempting to perform.

Baking may be a clear example of such active memory access. An individual has a desire to bake a cake from scratch. The individual may consult a cookbook in order to read the ingredients required to bake the cake, or if the individual is skilled enough, he or she may be able to recall the ingredients needed to bake the desired cake by accessing the memory files in their brain where the information regarding cake baking is stored.

But let's ask how does the average person, with an average brain, become an artist? Often an artist is considered to be an individual that holds a special talent that allows them to express an emotion or a meaning through some visual medium. A visual medium is anything that individuals can see and/or feel or hear which might include a painting, a sculpture, a filmmaker, a laser light show, sound generated by an instrument or group of instruments, writings by a novelist. Generally art is created with some intent in mind, such an intent might be to communicate a meaning by the artist to other individuals.

Still, how is the artist different than the average person? How does an artist 'have talent' for what her or she does, that is a different than other individuals? The idea that a person has a 'talent' for something, generally, means that they have a processing capacity in how their brain or body functions, superior to the average person. This superior brain or body function may be something that is recognized as a trainable trait, but is often regarded by the average person, as something the individual was born with.

If we consider that an 'artist' is born with a natural 'talent' or 'tendency' towards an art skill, then to be a 'prominent and/or successful artist' would suggest the individual's brain is somehow different than the brain of the average individual. This difference in the brain between an artist and the average individual may not be in the structure of the brain, but in 'how' the brain performs its duties.

To be an artist, it would seem that an individual would be able to gather information from their sensors regarding the features of the environment in which they live. The artist would need to store information in their memory applicable to their artwork. Artists tend to be considered 'visual' people thinking in terms of color, shapes, designs and being able to rearrange color, shapes and designs to suit their needs. In the case of such artists as a painter, sketch artist, sculptor, photographer or movie editor, they may have a stronger tie to the occipital lobes of their brain than the rest of the population (more hardwiring or nerve routes). That is, they may be able to take memory files and re-visualize the files in the occipital lobe or lobes of their brain. The occipital or rear portion of the brain is where the human brain is thought to see and analyze sight images detected by the eyes. Many people are able to remember images they have seen. Individuals with photographic memory capability are thought to be able to remember images they have seen in very fine detail.

If humans are therefore able to see images and remember images, then images must be stored in the brain as a form of a data files. Where the idea that we remember a 'name' is stored as a text file in the brain, 'images' our eyes 'see' must be stored in our brain as complex data image files. When we recall these image data files, we probably recall them from their storage site in memory, then transfer to the occipital lobes in the brain where an individual's subconscious and consciousness can review the image files. There are most likely more than one imagery area so that images can be compared.

In the case of a visual artist such as a painter, sketch artist, sculpture, or movie director, the data files must be capable to be moved to an area where they can be manipulated. That is, the data such as color, shape, size, and position must be able to be manipulated in the data file. Such a place in the brain may be referred to as a 'scratch pad' memory site, where data files are held for brief periods of time, and are a site where data files can be altered. A scratch pad memory site would be where entire files or selected portions of image files might be merged.

For instance, an artist might have an idea for painting mountain scenery. The artist might select the scenery based on a place the artist has visited. Inside the scratch pad memory the artist might manipulate the data file to change the mountain scenery from that of an afternoon picture to that of a sunset. The image is not complete, so the artist transfers the image of mountain scenery to long-term memory for storage. In his or her travels, the artist might encounter a moose. The artist might be so impressed with the physical beauty of the moose that the artist has a desire to place an image of the moose in the image of mountain scenery that is stored in an image data file in long-term memory.

The artist downloads the image data file of mountain scenery at near sunset to the scratchpad memory site. The artist then merges the image data file of the moose into the same scratch pad memory. The artist positions the moose in the scenery where the artist feels the image of the moose would be most appealing. The artist then stores the composite image back into long-term memory. When the time and medium are available, the artist downloads the composite image file back to the occipital lobe of the brain so that the artist can review or see the composite image in their brain. The artist then directs their output resources to reproduce the imagery by whatever output medium the artist is most comfortable with regarding the particular project, whether the output would be coordinating the muscles of the hands to produce a sketch, painting, sculpturing, translating the imagery into spoken or written language, or by regenerating the imagery by means of a photographic or computer assisted special effects technique.

Being talented or gifted as an artist, most likely means that several unique physical qualities exist. First, the individual's input data can be collected and stored in a clear and precise manner. Second, the individual possesses an area in the occipital lobes where large image data files can be viewed and reviewed. Third, the individual can store in the scratch pad memory, and in long-term memory large, intact image data files. The larger the bulk of information that can be stored and transferred, the more detail the imagery data files will have. The more detail, the more color options and three-dimensional quality the stored images will possess. The larger the data files and the larger the scratch pad and long term memory sites, the more images can be merged to produce final composite imagery. Fourth, an artist would have to possess large information buses in order to be able to transfer large blocks of information quickly between the occipital lobes, and the memory sites. Fifth, an artist needs to possess the capacity of being able to direct their output resources in a manner that produces a replica of the composite image file stored in the artist's brain, to the artist's satisfaction.

The work of most famous artists is thought to stimulate within the general public a message or an emotion when they view the artist's final product. Artists are viewed as being individual's that must 'feel' or 'sense' their surrounding environment, and transfer the emotional tie associated with their sense of their surroundings to their artwork. An artist is considered 'good' at what they do, if their artwork is capable of reflecting to the general observer of the work the same emotion the artist intended the artwork to exemplify.

Within the artist's brain function, the creation of composite imagery in the scratch pad memory site may be linked to the empathy or emotion unit located in the lower frontal lobe of the brain. As the artist generates a composite image, a sense of pleasure generated by the empathy unit may shape the composite image in one direction or another. If an artist painting a picture of a moose standing in the midst of mountain scenery is undertaking the task of generating the project because the artist is struck by the awesome natural beauty of a moose, then the moose may be the prominent figure in the final product of the composite image. If, on the other hand, the artist is more impressed with the appearance of breath taking colors that occur at the time the sun sets over a mountain range, the imagery of the moose inserted in the composite image file may be of only slight detail and relatively insignificant size and prominence in the final product. The moose would appear in some fashion in both final art products, but convey two totally different meanings and emotions.

The musician would seem to have a similar brain function as the artist, this may be why musicians are most often referred to as artists. Instead of using the occipital lobes as a canvas on which to draw, the musician uses Wernicke's and the Broca areas of hearing and speech recognition located in the temporal and frontal lobes of the brain respectfully. A musician, who creates new music, is able to string together musical notes and replay them in their head. They are able to hear the notes by replaying the memory files in the speech recognition portion of their brain.

So, if a musician is creating a new piece of music he or she begins stringing notes together. The musician sets up a memory file comprised of this string of notes. When the musician wishes to work on the file he or she transfers the memory file to their scratch pad memory site. In the scratch pad memory site, the musician adds to or deletes portions of the memory file. When the musician wishes to listen to the sound file he or she transfers the file to Wernicke's speech recognition area of the brain and plays the data file. The musician's brain hears the sound, not as real sound being generated, but as what it would sound like if an instrument or group of instruments were playing the sounds that comprise the data file. Like sight imagery can be interpreted by the brain, sound can be interpreted and re-interpreted.

The sound data file may become tied to an experience, such as the memory of a person, an event, or an emotion. A memory sound file may be related to an image file. So as the acoustic data string is being generated, an image file may be downloaded to the occipital lobe of the brain, and in the musician's mind, the image file and the sound data file may be linked. The empathy unit may send feedback, in the form of an emotion, to the conscious brain as the data string is being generated to help shape the outcome of the data string and therefore the resultant sound. As the data string is being constructed, the finalized pieces of the file can be translated into written word and documented, sung or hummed, or the hands can play the musical files on an instrument. The data string can be built upon until the length and arrangement of notes satisfies the musician.

What about the engineering mind? It is believed how an engineer approaches a problem is different than the average person, and quite different than how an artist would approach a problem (described above).

The question becomes, is the method that an engineer would employ to approach a problem truly different than the method an artist would use to approach a problem. If our brains are made the same, then shouldn't they work in the same manner?

An engineer's brain is comprised of a consciousness and a subconscious, the memory and the scratch pad memory unit just like everyone else. Like everyone else, an engineer also has a mathematics unit used to make mathematical calculations. This unit may also be able to store number files and be able to arrange number files into tables. The engineer's scratch pad memory site is most likely large, able to store large amounts of information. The ability to accommodate large amounts of information provides the ability to work with data files of three-dimensional objects.

The engineer's brain most likely has large information buses to transfer large volumes of data between the scratch pad memory site, long-term memory and the occipital lobes. Transferring large volumes of information allows the engineer to think in terms of three-dimensional objects. The engineer can retrieve a complex three-dimensional image files from long term memory and dump it into the scratch pad memory site. In the scratch pad memory, the three dimensional object data file can be manipulated. Colors can be changed. Object shapes can be altered. Object parts can be changed or deleted. Other files can be merged with the data file in the scratch pad memory. The engineer can then transfer the three dimensional image file to the occipital lobes of the brain and view the image of the data file in the occipital lobe. That is, when the data file is transferred to the occipital lobes, the engineer can appreciate what the three dimensional object image would appear to be as if the data file were indeed a physical object and mentally be able to reposition the objects so as to see the object from different perspectives (rotate the image in space).

In addition to specific information buses being larger in size, therefore, their brain's being able to transfer large quantities of information around their brains in different manners, the artist and the engineer may differ in the fact that they are loaded with different fundamental software programs. An engineer's brain may have software in his or her brain that better allows pattern recognition and spreadsheet table-type analysis than the artist, that may be equipped with software in his or her brain, that helps the artist to appreciate the details in objects and manipulate the objects and reproduce the objects in extraordinary detail. Another way to consider this would be to have two individuals with similar computer hardware. If one individual's computer was loaded with a spreadsheet software and the other individual's computer was loaded with computer aided graphics package, the first individual would be better equipped to analyze facts and figures, while the second individual would be better equipped to create professional quality drawings.

The aptitude any one individual senses in their life may in part be related to: (1) how their brain is constructed, (2) how the intricate parts of the brain are connected, and (3) the kind of software that is loaded as the primary programming in their brain. One of the objectives of this book is to bring to light the possibility that this exists, so that anyone and everyone can appreciate the fundamentals of how their brain functions so that they can get the most out of theirs.

On the other hand, it may be possible to stimulate the physical opening of new data buses or nerve connections, increasing data bus data flow to areas of the brain, or accessing dormant brain software, or add new software through education if an individual toils hard enough.

Training the human computer to perform desired or needed functions, may lead to increased efficiency or creativity as an engineer or artist or musician or other career path.

Unlike the desktop computer, the human brain is made up of living cellular units. These nerve cell units may undergo physical changes if stimulated properly to increase neuro-connections in a person's neuro network. Therefore, the brain may be able to adapt itself to the requirements of the tasks that an individual's brain is called on to do. An engineer may be able to evolve into an artist, or a musician or vise versa. The quality of being able to appreciate experiences or problem solving using a variety of perspectives, creates the individual, and a society, capable of accomplishing astonishing feats.

VI. WELL BEYOND THE REACH OF EVERY DAY FICTION

Φ
PULLING OFF THE
PERFECT CRIME

The Caribbean sea radiated with such a spectacular vibrant blue green color that was unmatched anywhere else in the world.

A voice spoke out over the dull roar of the crowd, "David Thomas…Message for David Thomas."

A red haired woman nudged her companion. A pair of stylish dark sunglasses sat perfectly on her petite nose. Spitefully she whispered, "Hey, someone's calling your name."

David Thomas, a well built eighteen year old college student stirred. Dark sunglasses braced the young man's attractive features. David sat up in his chair and waved his hand in the air to signal the resort courier.

The redhead chirped out, "Nevin,…Ah, David,…You're flipping him the bird." She rolled her eyes to the top of her head as if she were disgusted with her partner.

David looked up at his hand and realized his hand was up in the air, but only the middle finger was standing at attention. He immediately raised the other three fingers. He blurted out, "I'm still having some trouble coordinating some of the muscle movements."

Under her breath the redhead muttered in disgust, "Ya and if you don't get some other thing coordinated, sex on this trip might be better flying solo."

The hotel courier spotted the man with his hand in the air, and ushered over to the beach chair. The courier was smartly dressed in the resort's pressed white shirt and royal blue shorts. David Thomas said, "Yes what is it?"

The courier asked, "Are you David Zachary Thomas?"

David responded, "Yes, the one and only."

The courier asked, "Sir,…For security purposes can I ask you your room number?"

Without hesitation David responded, "The Governor's suite."

Satisfied, the courier pulled a letter from a side pouch and handed it to the resort guest.

The hotel was so plush, the courier did not wait for a tip. Tips were not expected. The courier disappeared.

The letter was addressed *To: David Zachary Thomas* and was marked confidential.

The redhead asked, "Who's it from?"

David scanned the return address. It was that of the law offices hired by the late Nevin Thomas. David said, "My lawyer."

David tore open the envelope. He pulled the letter from the depths of the envelope and quickly scanned the contents.

Inpatient, his companion stated, "Well David,...What does it say?"

It says, "Dear David...We are sorry to inform you that your uncle Nevin Alexander Thomas has recently died...A life insurance policy was worth ten million dollars...But since the untimely death was the result of an accident,...the accidental death clause in the policy multiplies the death benefit by a factor of five...Since your uncle was killed by a city bus, the city has proposed to generously award his estate with a three million dollar compensation package to terminate any further legal action against the city...Which I, being the executor of the estate, have accepted. The total sum of fifty three million dollars, less my fees, has been transferred to your Swiss Bank account...Best of luck."

"Sounds like you've won," remarked the sassy redhead with soft facial features as she squirmed excitedly in her beach chair.

"Better,...Didn't expect the three million from the city...to be that easy." he added, "I was on the verge of being broke as Nevin Thomas,...Now I'm unbelievably rich as David Thomas."

"How long do you think you can live off of fifty-two plus million?"

"Long enough."

"Think anyone will ever figure it out?"

"Not in our life-time...You set fire to all the evidence. Dr. Stephens was so secretive about his work, he didn't tell anyone,...He thought we was slated for a Nobel Prize. It'll take another hundred years for anyone to match Dr. Stephens' understanding of the brain's software principles he had discovered."

"Didn't you get the files?"

"Yep, downloaded his computer files to disc while he wasn't looking...They're locked safely away in a vault back in the States." David Thomas picked up a cigar. He bit the end off the thick, authentic Cuban fireweed. The sour taste of tobacco touched his lips. Striking a match, he lit the end of the cigar, then heartily puffed on the stogy. A broad, satisfied smile stretched from ear to ear.

The redhead smiled as she jubilantly lifted her half empty glass. With a sizzle in her voice she said, "To the perfect crime."

David Thomas lifted his half full glass and triumphantly responded, "To getting away with the perfect crime...having a second lifetime and a mountain of cash to enjoy it."

The redhead remarked, "You really should give up smoking."

"Naw, now I've got another forty years to enjoy it,...and you never know...at the rate I'm going...maybe many more after that."

36

REPROGRAMMING THE HUMAN COMPUTER TO CURE CANCER

Computers become old and obsolete, get thrown in the trash, they don't develop cancerous transistors.

Humans, unfortunately, do develop cancer. A particular cell, whether it be a lung cell, liver cell, stomach cell, or a blood cell can begin growing and dividing faster than normal. The reason why a cell might grow and divide faster than normal is still a mystery (though environmental factors and genetics have often been thought to be contributors). If a cell's metabolic rate is faster than its neighboring cells, the cancerous cell will steal resources intended for its neighbors, and invade into the space occupied by the neighboring cells. When such occurs to the extent that it can be physically seen or felt by the patient, or detected by a radiographic (x-ray) study, then it is usually considered a cancerous tumor. The word 'tumor' actually refers to any mass whether it be a benign or a cancerous collection of cells. Benign tumors are normal cells that grow in one location. Cancerous cells grow in one location, but can spread to other areas of the body and begin to grow in many locations.

To revisit the topic of the human defenses, again, there are two major cell lines. The T-cells are the roving police and the B-cells are the line of cells that produce the antibodies. Antibodies are like programmed cruise missiles designed to travel through the blood stream to all corners of the human body, and seek out invading pathogens commonly known as bugs—often referred to as an infection. The antibodies produced by the B-cell line, assists the T-cell line, by attaching the exterior wall of invading bacteria, viruses or parasites to make it easier for the T-cells to locate, attack and kill the pathogens.

Amongst the subsets of T-cells include the neutrophils for hunting and killing bacteria, lymphocytes for killing viruses, eosinophils for defense against parasites. There is a poorly understood subset of T-Cells referred to as natural killer cells.

Natural killer cells are thought to act in a surveillance capacity searching the body for cancerous cells. When they find a malignant cell they are suppose to destroy such cells. The frequency and wide variety of cancers which are known to invade the human body, may lead one to consider that the natural killer cells aren't very effective in performing their task.

The issue is that the need for natural killer cells has only become important over the last one hundred years of humanity's two million years of existence. Until the late eighteen hundreds, the life span of the average human was eighteen years of age. In nineteen forty-four, the average life span in the United States had risen to forty-eight years of age. In the nineteen sixties, the average life span increased to sixty-five. Currently, the average human in the United States is expected to live to the late seventies, early eighties.

Prior to the last fifty years, diseases tended to kill off the population. Urbanization and antibiotics has led to the extension of the life span of the average human. The extension of the average human's life, beyond the forties and fifties, has lead to many cancers becoming an increasingly more important factor regarding survival.

The neutrophils, lymphocytes and eosinophils would not be as effective in defending the human body, if the body's own antibodies were not available to assist them in their function of tracking down the adversaries. In today's medicine, when the body's defenses aren't sufficient to ward off a biologic adversary, a healthcare provider may administer antibiotics to the infected patient. Most antibiotics are designed to treat bacterial infections and some parasites. As years of research directed against fighting the common cold, and the deadly AIDS virus and lately, the West Nile virus has demonstrated, our capacity to synthesize effective antibiotics against viruses is nearly non-existent.

Still, if we could either communicate and reprogram the human body's B-cell line, or expand our synthesis of synthetic antibodies, we might be able to assist the body's lymphocytes in killing viruses and able to stimulate the body's natural killer T-cells to do a better job at seeking out and eliminating cancer cells.

Natural killer cells, like other T-cells, have the freedom of traveling to any part of the human body. If they could be programmed to seek out and destroy a particular tumor cells by coating the tumor cells with specific antibodies, cancers might be able to be attacked, neutralized, and eliminated from the body without need for surgery, chemotherapy or radiation therapy.

◆   ◆   ◆

The science of engineering cancer fighting antibiotics could usher in a whole new age of medicine. Antibodies or antibiotics to treat cancer.

◆   ◆   ◆

But lets not unleash a horde of Natural Killer cells just yet. What if we thought about cancer in a different light, a useful light? How could any cancer be useful in any way, shape, or form you might ask?

Well for a moment, let's think about the phenomenon of cancer in a way that an engineer, or any other analyst, might consider.

If you haven't been convinced up until now, that the human brain and the human body are like a computer, let me introduce to you the term *apoptosis*. Apoptosis arrived in the medical literature several years ago. *Apoptosis* means *programmed cell death*.

Apoptosis is the medical world's explanation for the aging process. The theory is that we age, and eventually die, because human tissue cells are programmed to die. There is a definite life span to our cells. There are only a certain number of times that human cells can divide into new cells before they wear out and die. It's analogous to going to your copying machine and making copies of a single drawing. As you made a copy, then take the copy and copy it, and then take the copy and copy it, eventually the details of subsequent copies would begin to fade and the subsequent copies would not look nearly as sharp and clear as the original. Well, after so many copies are made of a person's tissue, the copies become faulty, and at a certain time, the copied cells can no longer be copied and cell lines become disrupted. When a critical mass of cell lines can no longer be generated, because the copies are riddled with mistakes in the DNA, an organ or organs of the human body succumbs to the effects of aging. Aged organs become a focus of dysfunction or become a target of an infection or a cancer.

Well let's think about the concept of cancer again. Cancers represent a cell line that has evolved to grow more aggressively, and with more vigor and vitality than neighboring cells. What causes cancers cells to grow more aggressively? If we learned how normal human cells became cancerous, could such information be used to prevent normal human cells from becoming cancerous? Better, could we learn how to reprogram normal human cells to remain vigorous and exhibit vitality, so as not to age? In addition, could one consider possibly revitalizing old cells and making them young and vigorous again? The elixir of life, might be within the grasp of the new age of human computer programmers...the *neurocomputerologists*.

There are still so many secrets regarding the human body, left to explore and potentially utilize to help humanity stay healthy, and remain youthful.

37

DOES A Y2K-LIKE BUG THREATEN THE HUMAN COMPUTER?

The Y2000 bug referred to a problem that some computers had with the turnover of the Julian calendar from the year 1999 to the year 2000. Programmers tend to be people that design with efficiency in mind. One of the ways to be efficient in programming in the 1970's, 1980's, and possibly the early 1990's was to represent a year in 'two numbers', instead of 'four numbers'. Therefore, the year 1989, would simply be represented as the last two digits '89'. This truncation of the year works well until the years advance to the point one century turns over to the next century.

When the year becomes the next century, then the two-digit number becomes smaller than that of the current year. That is the year 2000, would be represented as '00'. In the case of trying to determine differences in years when larger numbers are subtracted from smaller numbers, negative numbers are produced and this can be confusing for both the computer and the user.

An oversimplified example would be a person started as a subscriber to the local newspaper in 1992. If in the year 1999, the sales manager of the paper wanted to know how many years this particular patron had subscribed to his newspaper, the sales manager could employ a computer utilizing a simple program to figure out the length of time the patron had been a subscriber. Most computers would have taken the year 1992 and truncated it to '92' and the year 1999, and truncated

314

it to '99'. Therefore, to solve this simple problem of determining how many years the patron had subscribed to the newspaper, the computer program would subtract '92', representing the year '1992', from '99' which would represent the year '1999' to arrive at an answer of seven years.

If, on the other hand, you wait twelve months to pose this same question, then you would be faced with taking the year the patron started subscribing, which would be '1992', truncated to '92' and subtracting this from the year '2000' which would be truncated to '00'. Now trying to perform the same mathematical calculation results in trying to subtract '92' from '00'. Since the result of this subtraction is a negative number or 'negative ninety-two' this could have either generated an 'error message' in a computer that is unable to accept negative numbers as a 'date' or it could have resulted in shutting the computer down completely.

If you were running the billing department of the newspaper and your computer hardware and software were not ready for the turn of the century, you might have taken immediate notice of the Y2000 problem, on January 4th, the first business day following the 1st of the year. If you had computers that used only the last two digits of the year and not all four digits of a year, then on the first business day in the year 2000, the newspaper might have sent out bills to its customers and the billing might have been figured out as follows: the computer might have figured that a patron that subscribed to the paper from January 4th 1999, till present and was overdue in payment on January 4th, 2000, could be billed for '1-4-00' subtracting '1-4-99' which might have produced either an indiscernible negative number that the computer would not have accepted, and therefore produced an error message; or the computer might have just shut down; or the billing computer might have taken the absolute number of the subtraction process '00' minus '99', which would result in a 'negative 99', and generate a bill that charged the patron for ninety-nine years of newspaper delivery.

Luckily, with the attention the turn of the century brought, most computer glitches were anticipated and solved before they happened. Anybody receive a bill for '99' years of something?

A HUMAN DESIGN FAULT?

Does the human brain have a similar fault in its design? January 1, 2000 A.D. represented a day in the history of the Earth. There have been many different calendars used by populations inhabiting the Earth. The Chinese, which comprise nearly a quarter of the Earth's population, use a different calendar system than the Julian calendar used in the west. To the Chinese calendar, our day January 1, 2000, had no specific meaning other than we were celebrating a day in history, which they obliged by participating in a portion of the celebration.

It is obvious that the human brain was not effected by the calendar advancing to the year 2000. But one should ask if there does exist a parallel defect in our brains, that may not be specific to the year 2000 itself.

BRAIN CAPACITY

In medical school, students memorized enormous volumes of material. To understand medicine, a student had to learn the language of medicine, as well as the lessons of the material. In the four years of medical school, one masters an entirely different language both in its written and spoken forms, and one assimilates the volumes of information that this new medical language conveys into a set of diagnostic tools.

Some students in medical school sincerely felt that the volumes of information required to be memorized were too overwhelming. I remember the class's mentor stressing to us that our brain capacity could indeed handle the enormous volumes of information that we were quickly trying to make our brains remember. Though we certainly did not memorize everything that we tried to learn, we did successfully memorize much of the information taught to us in those four years of intense learning. So, that if modern medical school training is any gauge of the ability to learn a tremendous amount of information in a short period of time, it would seem that the human brain has an infinite capacity to learn. There does not appear to be a limit as to what our individual memory can store.

ADDICTION

On the other side of the human experience, addiction to items in our environment seem to cause a breakdown in brain function. Addiction is most commonly recognized as a need to ingest illicit drugs to produce a modulation of brain function. The task of obtaining and then using illicit drugs can become all-consuming with regard to one's resources and time. Obviously, for a portion of the population seriously hooked on the use of such drugs, pursuit of crime becomes the only avenue by which a person may be able to satisfy their habit. Other items in the environment are also addicting. The use of cigarettes, alcohol, coffee can result in an addictive pattern of consumption. For others, the accumulation of wealth, the purchasing of precious coins, precious art, diamonds, jewelry may be an addiction. We as humans, often have a desire to possess things that exist in our environment, and sometimes this need to possess an item or items is conducted in what might be considered an addictive pattern (sometimes considered to be eccentric).

Such an addictive pattern of human nature, would suggest that biologic subroutines exist in the human brain capable of overriding normal human function, and in some cases, override human functions vital to the maintenance of the human body. That is, an individual that has succumbed to an alcohol addiction, may not consume food in proper quantities or proper nutritional content to meet the needs of the body because they are so invested in the pursuit of obtaining their next alcoholic beverage.

To account for the presence of such subroutines, it seems that there would be the existence of a coordinated effort between (1) a particular target file the body has identified that it wishes to pursue (money, jewelry, drugs), and (2) the pleasure center in the body. Once the target is achieved, the pleasure center triggers but the satiety center does *not* activate to down-regulate further pursuit of the target. The addiction subroutine would preferentially take charge of the body's physical functions such as walking, talking, grasping and combine these with eyesight that is tied into searching for the target of the acquisition whether it be drugs, money, or other item the brain feels a necessity to acquire and continues to feel a need to acquire.

Before we generate a negative feeling towards the existence of such addictive biologic subroutines in our brains, we must first consider the perils faced by primitive man. Primitive man needed to eat and cloth himself and his family. Both the food and the clothing often came not from the store on the corner, but from animals roving the environment where primitive man lived. Primitive man did not go to the local grocery store, select what meat he or she wanted, slap it down on the check out counter, pay the cashier and exist with the meat in hand. Primitive man first needed identify what in the environment was edible. Then he had to select which animal was edible that he might possess the capacity to capture and kill. Once the type of prey was selected, be it a fish, a bird, a deer, a wooly mammoth or any other target, the man had to determine the most energy efficient means of capturing and killing the animal. Decisions on what to hunt, and how to proceed with a hunt, were important, since there was only so much energy an individual could expend between meals. If an individual spent too much time hunting without capturing an adequate prey, the hunter and his family may have become too weak to carry on their existence. Obviously, once the prey was caught and the man and his family ate their next meal and possibly clothed themselves with the animal's skin, the man experienced a feeling of satisfaction, however brief that might have lasted.

All of this seems to resemble the subroutine described in addiction. That is, when primitive man needed to consider capturing and killing prey for food and clothing, something inside had to signal to him that the function of capturing food was needed. The satiety center in the brain is responsible for alerting humans that they are hungry and require nourishment. When the satiety center activates, the human starts searching for food. Once food is found and ingested there is a pleasure that the human feels in having accomplished the task.

This above scenario is reminiscent of what an addiction subroutine would act like. The addiction subroutine may override body function subroutines, depending upon how much the target of the addiction stimulates the human pleasure center and causes a release of internal endorphins. Internal endorphins are chemicals generated by the body that cause pleasure in the body by stimulating pleasure centers. In the case of addictive drugs, the drug itself may cause a stimulation of the production of internal endorphins, thus causing an independent pleasure stimulation at one or more pleasure centers in the body. The sense of pleasure can be a strong motivator. Overwhelming stimulation of pleasure can, on the other hand, disable an individual's ability to make rational judgments.

PHOBIAS

Alternately, biologic subroutines may warn the human body of danger in an attempt to prevent harm to the body. In the case of primitive and even modern man, if a person is confronted with a confining space or the edge of a cliff, or the edge of a window sill at a substantial height, a subroutine in the body may kick in and stop further forward action by the body to prevent the body from becoming stuck in a tight confining space or from falling from a dangerous height and becoming injured. Eyesight would recognize a target in the path of the human. This target would stimulate fear in the human. This generation of fear would slow or put a halt to the mechanics of the body to prevent injury.

Again, in defining possible failures in the human body's computer program, some members of the population suffer from compelling fears. When a fear becomes so intense that it results in disability, it is considered an anxiety disorder. Fear of tight spaces is commonly referred to as claustrophobia. Other morbid aversions include a fear of heights (acrophobia), fear of breaking mirror (catoptrophobia), fear of walking (basophobia), fear of bees (aphiphobia), fear of being locked in (clithrophobia), spiders (arachnophobia), fear of snakes (ophidiophobia), fear of anything new (kainophobia), abnormal fear of the male sex (androphobia). Many target objects or places can result in activating a fear subroutine. The reference I have lists two hundred different phobias.

These fear subroutines become pathologic, debilitating, when they interfere with normal body function. A woman that I had the opportunity to interview one time had a fear of rain (ombrophobia). When the sky was clear she could leave her home and go shopping. When there were clouds present in the sky, the woman could not leave her house, her fear was too intense. In essence, her human computer was shut down by the presence of clouds in the sky.

THE NERVOUS BREAKDOWN

Another form of failure in the human brain is what is commonly termed a 'nervous breakdown'. This may simply be that the person that experiences nervous breakdowns, has a subconscious that is unable to process the data inputs from the peripheral sensors in an orderly or timely fashion or the subconscious may be overwhelmed by too many conflicting interrupt signals. Control mechanisms usually transferred down from the conscious brain, may malfunction. When the subconscious mental processor is overloaded with data or conflicting interrupts, it may fail to function, similar to a computer that fails to function due to an error in the software program, an error caused by too many inputs occurring at the same time, or the occurrence of conflicting interrupt signals such that the CPU becomes overwhelmed and does not know which step to process next. Such an inability to process data or interrupt signals effectively, may be an inherent defect in a person's brain, or related to damage that may occur to the tissues comprising a person's brain.

CEREBRAL VASCULAR ACCIDENT

A stroke, otherwise called a 'cerebral vascular accident' (CVA) may make its appearance known by causing one or more motor and/or speech defects. A mini-stroke, often refers to a short-lived stroke, whereby the affected individual recovers quickly. A permanent stroke leaves an individual with one or more apparent obvious defect(s). A stroke is usually (a) the result of a lack of adequate blood perfusion to a part of the brain, or (b) the rupture of a blood vessel, causing an indiscriminant bleed to occur in a portion of the brain. Individuals that suffer a stroke, may recover their functional capacity, or they may be marred by a permanent physical dysfunction, emotional lability, personality changes, or cognitive deficits.

MID-LIFE CRISIS

Another form of apparent failure of the brain is what is commonly referred to as a 'mid-life crisis'. During such a crisis, an individual may make sudden, possibly drastic, changes to their normal routine or lifestyle. Often, such a mid-life crisis is seen as an adult who is reverting to a more childish behavior by rejecting or ignoring their responsibilities and obligations. Often such a crisis entails that an individual will leave their spouse and family. The person's actions, while the crisis is occurring, often includes leaving their job, possibly changing their career to a totally different work-effort altogether.

Such a 'mid-life crisis' may, in some, represent a personality data file that the individual had, that never before surfaced from archived memory to the conscious level of decision making. Let's take the woman who is married to a relatively attractive, stable man, and is the mother of three children; she leaves the security of her spouse and family to pursue a career as an aide worker in an unstable third world country. It is difficult on the surface to understand why an individual would reject the stability of a family and the comfort of living in the United States, to pursue helping people the individual does not know or has any ties to, and live in rustic surroundings, in the midst of possible political upheaval, violence and disease. When asked, the individual may say that it was their destiny, they were finally following a dream they have always known existed inside themselves. To follow the dream meant experiencing a feeling of satisfaction or possibly even relief. In some cases the individual may be correct. Inside all of us may exist a biologic software program that provides a template for our individual actions. When our lives follow the pre-existing template, we may feel satisfied, even comforted.

In some of us, we may not be able to pursue the template that exists within ourselves. Various social or learned interrupts may override the ambitions created by the career template in our brains. Expectations creating powerful interrupts generated by our parents, religious advisors, social rules or laws may block our template from its opportunity to be expressed, or in some instances, the opportunity just never occurred in the time-frame of a person's life.

A psychologist gave a series of presentations to our medical school class and in one particular presentation the psychologist recounted the story of a physician who ended up working as postman for the Postal Service. The story was such that this particular physician struggled through medical school, residency training, then started to practice medicine. Both parents died suddenly in an automobile accident. The physician had revealed to the psychiatrist that he had pursued a career in medicine based solely on his perceived need to satisfy the expectations of his parents. His father had been a doctor, so the family had groomed him to be a physician. Once the parents died, and the interrupt generated by the perceived obligation to the parents was extinguished. The individual very happily left the practice of medicine to pursue a job as a mail carrier, which he considered a dream job.

Errors may occur in our brains, in our personalities and affect, due to an inherent genetic defect in the brain, due to a stroke, due to situations where the sensory input overwhelms the processing power of an individual's brain, fatigue, lack of concentration, or due to the surfacing of a behavior data file that dominates the individual's life style.

AGGRESSIONS

Finally, considering that humans do have the capacity to turn upon other humans and do harm to fellow humans on a mass scale suggests that a defect in human biologic programming may exist. Roman rulers amassed large armies and conquered a significant portion of the European and Mediterranean theater. Attila the Hun led 700,000 barbarians on a rampage that stretched from Asia to Europe engulfing everything in their path. Napoleon, amassed the French armies and scoured Europe with his military might. The Spanish came to the New World and plundered the villages and cities of Central and South America. Adolf Hitler was able to control the masses in Germany and Austria and launched a military campaign that engulfed Europe on both Eastern and Western fronts. Stalin led the Russian people to put to death many thousands of their own.

It is curious how aggressive behavior and greed in humans can overshadow a person or a population's compassion towards fellow humans, to a point that harm and even death may be the end result of such aggression. Behavior that causes humans to set aside their empathy towards other humans, that causes humans to degrade fellow humans to a point that in an unprovoked manner fellow humans can be killed for sport or in the name of conquest; can this be telling us that their exists a flaw in our basic human programming? A flaw that can be exploited. A flaw in the human programming at the level of the subconscious, that others can use to take advantage of us as individuals, to manipulate our individual behavior to satisfy the manipulator's own agenda?

Unlike the Y2K bug, that could have caused computers to have ceased functioning, a flaw in human basic programming that facilitates humans to be manipulated in a manner that changes our behavior on a mass scale from peaceful to unforgivingly militant, is a far more dangerous computer software problem. Peace on Earth could therefore, possibly be achieved by identifying this biologic software error or glitch, and correcting it. On the other hand, not to search and identify such a biologic software problem, would suggest history will continue to repeat itself with each succeeding generation.

38

THE NOT-SO-OBVIOUS DIFFERENCE BETWEEN MEN & WOMEN

Beyond a possible synthesized woman's or man's voice, or possibly being dressed in a pink or blue exterior casing, desktop computers are asexual,...produced on an assembly line—enough said.

The sexuality of humans on the other hand involves extremely complex programming techniques, but may involve only a few simple protocols.

The not-so-obvious physical differences between the sexes dictate behavior of the sexes. Estrogen is a female hormone. Testosterone is the well-recognized male hormone. Estrogen is considered to provide women with their femininity. Testosterone is recognized as being the hormone responsible for masculinity. Estrogen is considered responsible for stimulating breast development, shaping an adult woman's feminine curvatures, the softness of her skin and the menstrual cycles a sexually mature woman experiences on a monthly basis. Testosterone is thought to influence the development of muscles in a man, the broad shoulder male form, facial and chest hair, the deep voice adult men have, and it is thought to be the catalyst most men have that interests the male in sex.

The not-so-obvious physical differences between men and woman are embedded in how the hormones are delivered to each of the sexes. Men derive much of their testosterone from their testicles (though it can be made in both sexes in the adrenal glands located in the abdomen above the kidneys). Men experience a somewhat constant supply of testosterone. Due to the relatively constant nature of testosterone in a male's system, the sexually mature male has a constant interest in pursuing a sexual encounter. This interest in sex may be seven days a week, twenty-four hours a day. Women often complain that men are too predictable. This predictable nature in a man is related to a relatively constant flow of testosterone through the man's system, which keeps him at an ever alert, ever heightened interest in sex.

Women on the on the other hand undergo a cycling of their hormones. Women secrete Follicular Stimulating Hormone (FSH) and Lutenizing Hormone (LH) in varying amounts from the pituitary glands in their brains. The levels of FSH and LH stimulate the production and release of estrogen and progesterone from a woman's ovaries. As a woman's monthly cycle begins, an egg begins to develop in a follicle in one of the two ovaries a woman carries in her pelvis. The follicle carrying an immature egg is nurtured by the levels of FSH and LH secreted from the pituitary gland. As the egg develops, the follicle that surrounds it produces estrogen. The estrogen causes the inner lining of the woman's uterus to grow, preparing it to accept a fertilized egg. When the single cell egg is fully mature, the follicle ruptures releasing the egg from the ovary. One of the true mysteries of nature is that the egg is able to successfully migrate from the ovary to the opening of the fallopian tube, the causeway to the uterus, without getting lost in the woman's pelvis. A fallopian tube exists on both sides of the uterus reaching out toward each ovary.

If the mature egg becomes fertilized by a man's sperm as it travels through the fallopian tube, then successfully embeds in the readied lining tissue of the uterus, the production of estrogen is maintained. If, on the other hand, the mature egg is released and there are no sperm or an inadequate amount of viable sperm present in the woman's uterus and fallopian tubes to successfully cause fertilization of the egg, then the egg passes through the fallopian tube, through the uterus and then out the cervix and the vagina as a discharge. If a fertilized egg does not embed itself in a woman's uterus, then the follicle that had nurtured the egg to maturation releases the hormone progesterone while decreasing production of estrogen. As the levels of estrogen diminish and the levels of progesterone increase, this causes the fertile tissue lining the inside of the woman's uterus to slough off. This sloughing of the uterine lining tissue is what the woman experiences as her monthly cycle often termed a 'period'. Men know this as the time of the month as the time when a woman bleeds.

The average interval length of time of a woman cycle is 28 days. This cycling interval may be longer or shorter in some women. The interval is usually regular, though it may be very irregular for some women. The cycling interval is sometimes affected by stress or illness, causing some women to lengthen or shorten the days that exist in a cycle. Some periods are skipped altogether in some women at varying times in their life. The most common reason a woman will skip a period is due to a pregnancy. Estrogen levels are maintained if a fertilized egg embeds in the uterine wall preventing the uterine lining from being sloughed off.

Ladies, no doubt about it, it is certainly a man's world,...that is until women act together in a coordinated manner to change the structure of the calendar. You ask why? Twenty-eight days is the length of the average woman's menstrual cycle. Twenty-eight conveniently divides into 365 (the length of one year), thirteen times with one day left over. If a woman had designed the calendar, possibly there would be thirteen months, with twelve of the months having twenty-eight days each, and the thirteenth month having twenty-nine days. Seems a thirteen month calendar would be much more convenient, instead of having women subjected to the irregular daily count caused by each month of the current calendar year as dictated by the Julian calendar. In addition, the Moon revolves around the Earth in a period of about 27 and 1/3 days. Other than the cloud of superstition that seems to surround the number 'thirteen', it would seem that a thirteen month calendar would be much more convenient and biologically satisfying.

As described above, midway through a woman's cycling interval, the estrogen levels are at their highest and a mature egg is released by a ruptured follicle for a five day journey down the fallopian tube to the uterus. It is in this five-day journey down the fallopian tube that the mature egg may be fertilized by a man's sperm, if the sperm are actively in the fallopian tube. Studies suggest that at this time, midway through a woman's cycle interval, she may subconsciously dress more provocatively and act more aggressive when around potential male partners.

So then exists the quandary of nature, which has set up the battle of the sexes for hundreds of thousands of years. The male of the species has a constant testosterone level floating through his system that makes him not only predictable, constantly vigilant for sexual opportunities, but also constantly susceptible to the advances of an interested woman. The hormones of the female of the species cycle, making the woman's sexual drive somewhat unpredictable to the woman herself and certainly unpredictable to a potential male partner. The swings in a woman's hormones, the periodic retention of water causing the feeling of bloating, the abdominal cramping causing fits of pain and discomfort all become factors that make a woman's mood unpredictable by either sex. The secretion of testosterone by a woman's body, throws an additional wildcard into the mix, as surges of testosterone create an additional unpredictable sexual appetite in the woman. Women may not understand the impact on a man of one day being very interested in sex, and another day being very uninterested in sex; though those women that do understand this vital concept, may recognize their infinite capacity to torment their mate.

So one of the most frequent battles between men and women occur in a couple when a woman rejects a man she is been sexually intimate with either in a marriage relationship or a courting relationship. The woman may be at a low portion of her monthly cycle, and have no interest in sex. Her male counterpart, on the other hand, nearly always

possesses an interest in sex. If the woman fights with the man or rejects
the man and he removes himself from her sphere of influence, and the
man encounters a woman who openly expresses a desire for a sexual
relationship, the man may succumb to the advances of the willing
female.

The primary woman will label the encounter of the male and the
secondary woman as a breach of their intimacy, but the fact that is
often overlooked is that the male has a constant hormonal drive to
engage in sex. The man becomes guilty for breaching the confidential-
ity of the primary relationship, but over the ages, little emphasis has
been placed on the woman's possible lack of partnership except at
times when the woman wishes to be a willing partner. The man, at
times, is a victim of the constant nature of his sexual drive. Not only
does the man have to divert the constant nature of his desire to engage
in sex, but he may be swayed by an alternate sexual partner when his
regular partner is unwilling and remains unwilling.

The battle of the sexes goes on day after day, year after year, century
after century, because we understand so little about ourselves. There
are many bloody wounds that have occurred over the ages due to these
poorly understood differences.

Then there must occur sporadic rises in estrogen levels in women.
Occasionally, a woman will suddenly reveal her interest in sex to a perspective mate. Such apparent illogical, aggressive sexual behavior on
the part of the woman, often *bewilders* both partners.

Our problem as a species, is that we often feel we are *above* Mother
Nature's long established protocols. If men are on a two to three day
cycle, and if women are on a twenty-eight day cycle, and if when
women are interested in sex, they are capable of accommodating multiple partners where men are capable of reaching orgasm once, maybe
twice; the formula is that of species survival, not social stability. In
terms of the human species, societies rise and fall, the laws that societies
establish to govern relationships between men and women come and
go, while survival and propagation of the species remains the ultimate

mandate. The directives of survival may not insure the proper rearing of offspring once they are born, but what it tries to insure is the birth of offspring. Despite how much we wish to write laws of marriage to govern the behavior of sexual partners, though it may work for many, there will remain the exceptions because Nature's Forces remain stronger than man's willpower.

We, unfortunately, understand so very little about the precious value in our sexual differences. We fight over these physical differences, instead of working with them. It seems that ideal dream of romance occurs only when, despite the physical obstacles imposed by the differences between the sexes, enough tolerance exists between two sexual partners that an ongoing relationship is nurtured by both partners and thus is able to flourish, which at its best might seem like a flowering cactus in the heart of an otherwise desolate, arid desert.

So humans have bestowed a tremendous quality to computers, they are asexual, and therefore do not have to devote any memory space or processing power toward seeking a soulmate.

39

UNLOCKING THE REAL DIFFERENCE BETWEEN MEN & WOMEN

Napoleon Bonaparte is arguably one of the greatest battlefield generals that ever lived. He controlled an army of over 80,000 French soldiers and marched on Europe, conquering much of the world as he knew it. Napoleon's testosterone level had to have been extremely high in order to have been able to control the lives of tens of thousands of men, directing thousands to their deaths on the battlefield. Napoleon had concubines. He told his wife Josephine that he loved her, and that his concubines meant *nothing* to him. This may have been so. Not to downplay the importance of feminism or the unity of marriage, but there are obviously real differences that exist between men and women, that involve more than the physical differences that we can see and what we are usually are capable of appreciating.

There are certainly many easily recognizable physical differences between men and women. There exists quite a different shape between the male and female body. There is a difference to the contour of the shoulders, torso, hips, and legs. The size of the hands and feet certainly are quite different. Obviously, there is a disparity in muscle size and distribution. It goes, almost without saying, that women have internal sex organs, the ovaries and the uterus that men do not possess. In addition, women possess mammary glands on their chest, which result in various sized breasts, that men do not generally possess.

The age-old mystery of why men and women inherently act differently may have a simple explanation. Most people acknowledge that the average boy and average girl engage in play activities that tend to be gender specific. That is, left to his own choices, a boy will pick up construction or military type toys and entertain himself with them. A young girl will pick up dolls and household toys, and mimic her mother.

In the past, we have shrugged this off as instinctual actions. Some experts have explained away such behavior as actions children pick up from observing their same sex parents. Most parents are amazed at how gender specific young children do choose their play toys and at such a young age they display these choices.

The explanation for such behavior may actually be very simple. The human brain is on average of similar construction and size whether it is a man's or woman's brain. Therefore, it is reasonable to say that there are limitations as to what can be crammed into the inner working of the human brain. These limitations to size of the human brain dictates that there must be choices as to the allotment of certain parts of the brain. Observing behavior, the difference between men and women would dictate that men possess more intricately connected logic units, and women have more effective empathy units, comprising the inner workings (via hardware and/or software) of their brains. On average, men therefore, might be better problem solvers. Women on average, might be better equipped at caring for the children and family and interacting with society. Therefore, in young children, when a boy reaches for a truck, tank, or bulldozer he is exercising his emerging capacity to solve problems and build structures. When a young girl reaches for a doll to play with, she is exercising her emerging capacity to care for future offspring and family. These distinct activities would suggest the sexes contain different pre-programmed memory files or programming features.

Another explanation for the differences between the sexes may be in how the brain is *hardwired*. It is conceivable that the general structures of the brain of men and women are the similar, but are the individual segments of the brain connected the same? Recent research would suggest that in a typical man's brain, the data buses that connect the front and back portions of the brain are larger in size and more active, than in a woman's brain. This would suggest the average man is able to focus on tasks more intently than the average woman. On the other hand, the in the typical woman's brain, she has larger and more active connections between the right and left hemispheres of the brain than the average man. This would suggest the average woman is able to concentrate on much greater sensory inputs, than the typical man, and that the average woman is able to command better verbal skills than the average man. Boy, doesn't that sound right on target!!!....How many men want to be just left alone to work on their tasks, and how many women are amazed that their male counterpart *never* sees the dust building up on the curtains, or that their male counterpart is relatively inept in social situations.

Certainly in many households, men are given the task of solving structural problems around the house, while women are charged with maintaining the appearance of the house. Actually, since Mother Nature's *prime directive* is creating balance in the universe, the separation of talents and tasks between the two sexes works well. Women identify problems and give them to the men. The men then solve these problems. The woman then congratulates her male counterpart on the completion of the task. The man feels secure about his role in the family. The woman then identifies further problems and charges the man with attending to these tasks. This is an endless cycle, but a balanced cycle, that results in survival, success, and prosperity of the family.

Problems occur if the woman fails to recognize that the tasks may be too difficult or costly for the man to perform, or if the woman does not adequately congratulate the man on the completion of the task. Further, problems can arise in the relationship if the man does not attend to the woman's tasks in a timely fashion, or completes them in a purposeless way without regard to the quality of the work.

It all boils down to the fact that on average, women and men probably have different allocations of rudimentary parts of their brains and processing capacities. Men, in general, may devote more processing power to the logic programming, which provides them with a greater capacity to solve problems whether mathematical, structural, or otherwise analytical. Women, in general, may devote more processing power to the empathy programming, which provides them with the all-important capacity to nurture and maintain relationships inside the family unit. She possesses the tools to solve social problems.

Variation is an important means of species survival and prosperity. All around us exist individuals and variation. It should not be construed that women cannot solve mathematical or structural problems. Some of the great analytical minds in history certainly have been women. It should also not be construed that men cannot be sensitive toward the emotional needs and well-being of the family unit.

On average, the sexes have an important distinction in how they think and how they approach problems. A difference that separates the sexes often leads to misunderstanding because the priority of logical thinking in the male may not coincide with the empathetic thinking of the female counterpart. Diversity is often a strength, and in this case important difference between the sexes may cement male and female relations together contributing to the survival of the species.

If in the world of primitive man, both man and the woman nurtured the children, and no one set off to find food or a means of clothing the family, the human race would not have survived. The fact that one mate hunted and the other mate protected the children, obviously led to role-playing, but also led to survival of the family unit and ultimately the human race.

40

THE SECRET BEHIND THE MOST POWERFUL FORCE IN THE UNIVERSE

Understanding the Human Computer could not be done in a justifiable manner without including a discussion regarding the role sex plays in the dynamic computer system. Sexual relationships entail some of the most complicated parts of our human existence. Sexual relations surround us whether we are single and actively searching for a partner, or married and maintaining a constructive relationship with a soulmate. Whether we are single or married, sex or fantasies about sex significantly impact our lives, influences our decisions, and drives our ambition on a daily basis.

Animals make sexual selection on the basis of fitness. The stronger the animal's fitness to survive, amongst the elements, and the stronger the animal's drive to mate, the more likely the animal will successfully mate and reproduce. This assures a fit gene pool, securing the species survival given current environmental forces. Ill, injured, or genetically inferior animals either die or fall prey to predators.

The human selection process most likely started out similar to other animals. Homo Erectus, primitive man, battled fierce animals on the treacherous planes of Africa to survive in the wild 1.8 million years ago. Selection of a mate was probably based almost solely on the survival of the fittest mates available who survived long enough to mate.

As technology has developed, and man has been able to rise above the dangers of his environment, the idea of primitive survival of the fittest has lost its necessity to act as the only selection process for mating.

Humans have been able to conquer the planet and for the most part eliminate the threat of animal predators. The set up of social structures and the ability to cultivate and accumulate wealth translates into an artificial measure of fitness. Humans care for and protect the ill, injured, and the disabled.

The human computer takes these factors into account when selecting a mate. Three levels of mating selection may exist:

(1) Function as a pure sexual attraction.
(2) Function of a supratentorial selection.
(3) Function on the basis of the mysterious, incredibly attractive force of TRUE LOVE.

These mating selections are described as follows:

(1) Pure Sexual Attraction

Pure sexual attraction functions as a means to insure mating is a simple concept. Men are attracted to women on the basis of the woman's shape, size of the woman's breasts, size of the woman's buttocks and abdomen, the willingness of the woman to engage in a sexual encounter. A woman is attracted to a man on the basis of the broadness of the man's chest, size of the man's arm muscles, firmness or size of the man's buttocks, the man's aggressiveness and willingness to pursue the woman for purposes of a sexual encounter. Other factors include hair color, personal hygiene, selection of clothing, and various perfumes and colognes may act as attractants to gain the initial interest of a potential mate. Both sexes often become susceptible to a sexual encounter on the basis of interest exhibited by the aggressor's expression of interest and persistence of their expression of interest. Sometimes though, neither elaborate expression or persistence results in a shared attraction by both parties.

(2) Supratentorial Selection

The function of the supratentorial selection is on the basis of a person recognizing that a sexual relationship with another person may lead to some form of social-economic gain or may be a matter of convenience. In the realm of convenience, one might marry because they feel that they are getting older and a person of the opposite sex that they have befriended seems to be a good candidate as a spouse, so they marry and mate. In the realm of social-economic gain, a person may recognize that a person of the opposite sex is attracted to them on the basis of primitive sexual attraction and that if they agree to the mating, they will share in the economic or social benefits that the other individual brings to the relationship.

A subset of the supratentorial selection could include the observation that people who seem to be opposites, appear to be attracted to each other when such individuals share close quarters. People may recognize that they have emotional, social or financial deficiencies. They may recognize that a potential sexual partner could fulfill a deficiency they possess.

Opposites Attract

As an example, a man or woman may have an introvert personality and may feel comfortable in their manner of life. The individual may recognize that their introverted life-style could be broadened. Interaction with an extraverted individual may cause the introverted person to develop a bonding to produce balance in the individual's life. The desire for bonding between the introverted individual and the extravert individual may supercede other sexual needs.

Close Living Quarters

Close living quarters can induce relations between the sexes. When the two sexes occupy close quarters, twenty-four hours a day, interest in individuals of the opposite sex increases. Close quarters cause individuals to focus on characteristics of individuals of the opposite sex. Conversation and close contact may occur in situations where it would not otherwise occur. This conversation and close contact may bridge barriers that normally exist between individuals of the opposite sex and create opportunities for interest to flourish and eventually facilitate the pursuit of an intimate relationship.

Prostitution

An obvious example of sex for gain is prostitution. A woman or man may make arrangements to meet a prospective sexual partner at a certain location. The arrangements may be made through a third party. The prostitute will remove their clothing and engage in a sexual act possibly without any predetermined selection criteria other than the price is right. The sexual act is performed based solely on a reward system, usually the exchange of money.

(3) True Love

True love is considered the quintessential time in one's life. True love ranks in the universe as an event in a person's life that has the fiery nature of galaxies colliding, the gravitational attraction of a black hole, has the mysterious beauty of a stellar nebula, shines with the intensity of a supernova, and has the unmatched brilliance of the quasars.

So, after 4.6 billion years, with billion upon billions of people experiencing it, what is TRUE LOVE?

The concept of true love may actually be a force that does indeed sweep us off our feet, radically changing our lives. True love may be the ultimate force in the universe as far as we know it. This ultimate attractive force between two people of the opposite sex may seem unpredictable to us, but may actually be very predictable to our subconscious computer. The *purpose* of this powerfully attractive force may be *a means to keep the human program intact.*

True love may simply be the facilitating mechanism that keeps the human DNA intact. When two sexually mature adults meet, each carrying like DNA that matches (that is *greater* than the average 99.9% pure), the two individuals sense an overwhelming desire to mate. Beyond the image file preprogrammed in an individual's brain that tells them what the opposite sex is suppose to look like, are audio files. The audio files may not be words, but sounds that cause a reaction. Wernicke's area of the brain may not only screen out sounds, and decipher language, but screen people we encounter for the purity of the genetic code they carry. Sounds generated by a perspective potential mate, might activate a memory file in an individual that tells the individual listening to the sound file that the person whom is speaking is their soulmate (or not).

Once a soulmate, that is a sexually mature adult carrying DNA that matches the listener's DNA, is identified, the listener may activate all available body and personal resources in order to attempt to secure a relationship with the identified soulmate. Social and learned interrupts may be down-regulated or completely nullified, by the more pressing need to pursue a mate carrying matching DNA code. Primary programming protocols may dictate: (1) survival of the individual, superceded by (2) survival of the species, superceded by (3) survival of the genetic code at *any and all* costs.

LOVE CONFLICTS

The elements of human conflict may occur when any of the three above-mentioned sexual attractions clash. Such conflicts may occur due to (1) decisions we make to legalize or otherwise toil to make permanent our relationships, (2) timing of our encounters with persons of the opposite sex, (3) responsibility to children, (4) responsibility to family, church and social structure, (5) relationships previously developed by the person whom we are attracted to, (6) the dominance of our primitive sex drive or our supratentorial brain.

Disastrous human conflicts may occur when one is mated with a partner that they chose solely on the basis of primitive sexual attraction or on the basis of social-economic gain, and they encounter a mate that they fall madly in love with due to the fact that the new mate carries genes which represent the other half of their match of the computer program and natural forces draw them together.

Other equally conflicting trouble occurs if one mates on the basis of convenience, social economic gain, and then encounters a mate that activates an overpowering primitive sexual attraction. Certainly, sexual relationships may be prone to manic highs and lows, spawning numerous break-ups and returns to the relationship. On the other hand, such sexual relationships may be very stable if the sexual intensity remains at a high level.

As environmental conditions improve in the modern, industrialized parts of the world relationships on the basis of true love may be very stable. Both partners enjoying an unexplained satisfaction being in the relationship. On the other hand, they can be very volatile if one or both partners have ongoing convictions either socially or with pre-existing sexual partners. These relationships can be fast and furious. They can result in dissolving an existing marriage. They can result in one partner entering an energized sexual relationship and terminating the pre-existing relationship.

Relationships involving true love might *not* be long lasting relationships as one would hope. In fact, they may be very brief in their existence—just long enough for mating to occur and both parties to feel satisfied that they have fulfilled their sexual destiny.

Let's take a vibrant young woman who has been selected to become the wife of a king. Let's say she is not at all attracted to the king.

Would she let her lack of interest in the man sway her from becoming queen of the land? Probably not. So the maiden marries the king out of convenience and becomes queen. Her king is twenty years her senior and does not possess the same sexual stamina as does the young queen. Wheeling her wealth and power, the young queen creates opportunity. She meets a young men of her own age and is attracted to him, but she does not infringe upon her wedding vows because the force of the attraction is not strong enough to risk being discovered to be an unfaithful wife and hung. In the queen's travels though, she does encounter true love. In this case, her sexual attraction for the man is categorically off the scale. She enters the sexual relationship. Once mating is complete the queen has a choice to make. The queen can: (1) hide the relationship from the king, (2) terminate the relationship, (3) leave the king for the young suitor. The queen elects to terminate the relationship once her sexual desire has been met, and if she were to feel threatened enough by the relationship, she may even have the young suitor terminated.

So let's re-visit the *True Love* concept again—basically, it refers to two people meeting, and their internal biologic programs determine, at the subconscious level, that they should mate for the good of the survival of the biologic program. At the conscious level, both individuals overcome all physical, social and governmental obstacles to be together, to mate and produce offspring.

At a romance novel level (1) The rich daughter of a oil barren, whose family estate is worth tens of billions of dollars, all of which could go to the only heir, their daughter, a daughter who instead, meets up with some deadbeat drifter, robs banks and ends up in jail for the rest of her life. (2) The wife of an investment banker decides to give up her furs, jewels, and expensive car, to run away with a gourmet chef who works at the country club her husband and she have a membership. (3) The faithful, young wife of a minister has an unexpected, per chance meeting with a tanned, very attractive ski instructor, who ensnares her to have a one time fling while at a New Year's Eve party after her spouse gets drunk and passes out.

For all the romance novels readers, it boils down to one of three central themes: (1) pure sexual attraction, (2) sex for gain, (3) or the biologic computer program found a match and come hell or high water, the program will see to it that the two individuals mate, so as to insure the survivability of the genetic program in its purest form. The social, religious, and/or moral and ethical outcomes the participants face, are inconsequential to the biologic program driving the life processes on the planet.

AT THE PHYSICAL LEVEL

How would all of this come about? Stringing together the components of the human brain may provide the answer. As discussed earlier in Chapter 29, locating a sexual partner or mate may simply involve a form of target acquisition. Target acquisition may be in two forms. One form of target acquisition may involve a revolving target. When one rises in the morning one might seek out clothes to wear. The target, in the target acquisition center, may switch from one article of clothing to another until the wardrobe is complete. After a person is dressed, the satiety center in the brain may stimulate the brain to search for food. The target center of the brain dials up images of food for the eyes to search for. After breakfast has been eaten, the brain may seek out the car so that the individual can travel to work. The target in the brain's target acquisition center may be that of a car that the individual must locate out in the parking lot outside their house or apartment.

Other target acquisition images may be more permanent in the brain. A man actively searching for a sexual partner may have already stored in his brain the image of the acceptable woman he is searching for. Whenever his eyes lock onto the image, or a like image, the man may become aroused to pursue the target. Thus a man may have the image of a woman with the dimensions of 36-24-36 imprinted in his mind, and may even have a certain hair color, skin color, skin texture and facial features imprinted in his mind. When the eyes come across a target that fits the image in the memory file stored in the man's brain, then the male hormone center is activated and a sexual arousal is stimulated.

Having an image of what the opposite sex appears to be stored in the brain, makes sense when one considers that in the animal kingdom it is not uncommon that the male and female of a species do not look alike. Often times, size and color of the male and female of a species may vary quite distinctly. So an animal may not be able to rely on looking at itself in order to find an appropriate mate. The animal must rely upon some other means to discern how an appropriate mate should look. There may be pheromones (a chemical released by one member of a species to influence the behavior of another), colors, movements, or voice signals that might attract one mate to another. Still, somewhere in the animal's brain there must be a data file that instructs the animal how it can expect the opposite sex to appear. This will cause the animal to lower their defensive guard and copulate with a member of the opposite sex of their species.

In the world of primitive man, there was obviously limited means available to an individual to know how the individual (himself or herself) appeared. Before plate glass and the mirror were invented, all that was available would be a reflection of one's image found in a pool of still water. Unlike today, primitive man probably did not have much sense of what he or she actually looked like.

Therefore, like other animals in the animal kingdom, the primitive human brain needed some preset image stored in the brain to stimulate an individual to find a sexual partner amongst the environment he or she lived. Additionally, having sex with a partner that was not sexually mature would not result in successful procreation of the species and having sex with a person of the same sex would likewise not result in successful procreation of the species. The human brain then, for necessity sake, required storing an image in a man's brain of an acceptable woman with whom to mate, and storing in a woman's brain, the image of an acceptable man.

For a man, the image of shapely curves, soft skin, well-defined breasts, tight buttocks, and the presence pubic hair, would act as a signal in the man's brain that the appropriate target had been acquired and that mating might be successful. For a woman, the image of a broad chest, enlarged arm and buttock muscles, erect posture and the presence of facial and pubic hair might signal in the woman's brain that the appropriate target had been acquired, and that mating might be successful. The idea of facial features may or may not have been a driving stimulus to primitive man. The invention and availability of the mirror, suggest the acceptance of the facial features of a mate based on how much the facial features of a potential mate look like the facial features of a person searching for a mate. After years of seeing our own face in a mirror, while we struggle to groom our appearance, we may become at ease with our appearance and, therefore, when a person of the opposite sex with facial features similar to ours comes into sight, we may find them attractive. Certainly, it is a well-known phenomenon that married couples begin to look very similar in the appearance the longer they are together. This is most likely a reflection that persons seek out mates that have similar facial features.

The image of a young woman may be the target of acquisition in a man's brain. When a man enters a room crowded with strangers of varying ages, his eyes automatically scan the room for the presence of sexually desirable women. Young, sexually immature girls are excluded because they lack the proper visual characteristics to register in the adult male's target acquisition center. The adult women in the room are quickly divided into two groups, those that meet the target acquisition characteristics stored in the man's brain and those that don't. The women that appear to be of sexually active age fall into the interest category. In the initial scan of the crowd those women, with soft skin, healthy looking hair, prominent visible breasts, tight buttocks, who stand erect may be the most attractive, because they may be the closest to the image set in the target acquisition file stored in the man's brain. Further inspection of the crowd may hone down the possible sexual partners available to the man as he traverses the crowded room, based on hair color, specific age, specific breast size, specific waist size that the man's brain is trying to match up to the target data file stored in his brain. A variety of social, religious, economic and self-esteem 'interrupt factors' may also play a role in the selection of a sexual partner.

There certainly exist sexual characteristics that cross over between the sexes; some men are more feminine in their appearance and mannerism, while some women are very masculine in their appearance and mannerism. The distinctive examples of this can be seen in the variation of the simple chromosome make-up. Humans possess twenty-six pairs of chromosomes. The pair of sex chromosomes are the only pair that are significantly different in gross appearance and they determine the sex of the individual. Men possess one 'x' chromosome and one 'y' chromosome. Women possess two 'x' chromosomes. In a condition of genetic variation termed Turner's syndrome, the individual has only one 'x' chromosome (no second 'x' chromosome or 'y' chromosome). Also referred to an x-chromosomal monosomy, these individuals with only one sex chromosome, present with a female phenotype, short stature, shield-like chest sexual infantilism, which occurs at a rate of 0.6 per 1000 live births. Patients with the genetic variation known as Klinefelter's syndrome, the individual possesses two 'x' chromosomes and a 'y' chromosome. These individuals who possess three sex chromosomes may have small firm testes and a varying form of gynecomastia (male breast enlargement).

If the physical genetics can vary, and the physical characteristics can vary, then it is a small stretch of the imagination to say that the primary image an individual will use in sexual target acquisition may cross over between the sexes due to genetic variation. That is, the majority of the men may have the image of a woman in their data file, the majority of the women may have a man in their data file. Some men may have a man in their target data file. Some women may have a woman in their target data file. Some persons may have both the image of a man and a woman in their target data file, thus being labeled AC/DC for interest in both or either sex. There may be individuals that have no target data acquisition file and therefore little to no interest in sex. Some of us may have two or more 'primary target acquisition files' in our heads, which could be one reason, why when dating, it can be so confusing if a man finds himself searching for more than one woman, or a woman finds herself searching for more than one 'perfect' man.

The Final Equation of Love

So let's take the chemistry of love, and boil it down to one single equation. So when a man selects a mate, he may have located the individual on the basis of: (1) attractive physical features, and/or (2) the wealth or position the individual possesses, and/or (3) with clear consciousness, he has *no* idea what has attracted him to the individual he has a desire to mate with. Women have the same equation, but due to the fact that rearing children is an exhaustive and demanding proposition, the woman might analyze the potential mate in the light of how stable the ultimate relationship will be.

41

I WOULD LIKE MY COMPUTER TO WORK MORE INTENSELY FOR ME

The high-speed Central Processing Unit in the personal computer, spends a significant amount of time sitting idle. The CPU twiddles its thumbs in the nanoseconds that it is waiting for other portions of the computer to send it either commands, or data to crunch in some mathematical formula.

The idea of *pipelining* is a conceptual effort to get the most out of a CPU by filling the empty time slots, where the CPU is sitting idle with nothing to do, with tasks. Still, while you're typing in your next command to direct what you want the CPU to do, the CPU is sitting idle. Further, when I leave my computer to explore other avenues of my life, my computer really does sit idle, waiting anxiously for the next keystroke to be depressed on the keyboard, or the next movement of the mouse or flash of an infrared beam.

It would be nice to be able to give the computer extra tasks to work on, either while I was using the computer as I normally do, or while I'm away. It would be great if the computer would work on tasks I assign it, until I decide to come back and check on the efforts of the computer in five minutes, an hour, a day, a week or even a month later.

I give my brain problems to ponder, and my brain gives me the answer to my questions in five minutes, an hour, several hours, days or even years later. It would be nice to be able to task my computer, like I do my brain, and let the computer try to figure out some of my life's more tedious or difficult problems…while I'm out having fun!

42

ORIGIN OF THE HUMAN COMPUTER THE <u>INVERSE DARWIN</u> THEORY

The origin of the desktop computer started as the product of numerous ingenious ideas crafted into crude boxes by men and women molding visions into reality. The desktop computer has undergone a tremendous growth spurt over the course of the past twenty years, and continues to evolve almost on a daily basis.

Every human life starts as the product of two sex cells. One sex cell is provided by the mother, referred to as an egg. The other sex cell is donated by the father, referred to as the sperm. Each sex cell carries half the DNA required to successfully construct a human, or twenty-three chromosomes. When the two sex cells fuse, they share their DNA and create a new human life by fusing the forty-six individual chromosomes they each carry into twenty-three pairs of chromosomes. The cell produced by the combination of the two sex cells is referred to as a zygote. The zygote carries the reconstituted DNA. All human cells, except for the sex cells, carry twenty-three pairs of chromosomes in the nucleus of the cell.

The sex of the individual is determined by the chromosomal makeup of the twenty-third pair of chromosomes. The individual's mother always donates an 'x' chromosome to the individual's genetic code. It is the father that in the twenty-third chromosome, either donates another 'x' chromosome, that results in a female or the father donates a 'y' chromosome, which results in a male offspring. So women carry twenty-two non-sex chromosome pairs and a pair of 'x' chromosomes. Men carry twenty-two non-sex chromosomes and in the last pair an 'x' and a 'y' chromosome.

Deoxyribonucleic acid (DNA), comprised of three billion base pairs of data, act as the blueprints of the human body. In humans, the DNA is arranged into twenty-six separate double helix strands referred to as chromosomes. The DNA blueprints act to create all the structures that comprise the human body.

Two sex cells, each carrying half of the structural blueprints for the human body, fuse and after much division and differentiation create the complex multi-celled, multi-organ, biped human body. This intricate, nine month process, has occurred relatively accurately, replicated billions upon billions of times over the span of possibly two million years. Is it therefore, that much of a stretch of the imagination, that all life on the planet shares a similar process?

It is known, that though different races of humans exhibit different physical appearances, all humans share almost an identical DNA structure. As discussed earlier, human genetic coding is 99.9% pure. It is further known that humans and monkeys share nearly identical DNA. Humans are not thought to have evolved from monkeys, but that monkeys and humans represent different offshoots from the same line of chromosome material. A stretch of the imagination is that the banana shares nearly 50% of its DNA coding with Humans. Do we really have that much in common with a banana?

Recently, the idea of evolution has taken a new, very interesting twist. It is now thought that the animal life on Planet Earth is the product of a similar set of chromosome material. The Hox genes are thought to determine the head to tail anatomy of many animal species. Insects and other invertebrates have a single set of Hox genes, while vertebrates (animals with a spine) have four sets of Hox genes. The

Hox genes in the vertebrates are nearly identical to that found in insects. There exists in the genetic coding these important master Hox genes that turn on and off, possibly dependent upon environmental factors. When certain genes turn on, the expression of a physical characteristic occurs. It is currently thought that fish, legged terrain animals and birds all carry similar genes, but their bodies are the product of the expression of different sets of master genes being turned on and off.

Over the 3.85 billion years life has existed on the face of the planet, it has been thought Mother Nature has experimented with many different forms of life. Mother Nature's experiments have been haphazard, and have produced numerous variations of life, some successful, other forms have become extinct relatively quickly; some extinctions being the result of changes in environmental factors that the biologic organism could not adequately adapt fast enough in order to effect survival. It has been speculated that the genetic expressions that have lead to the development of the human race is a product of chance, coincidence, luck, and just being in the right place at the right time.

Thinking that Homo Sapiens, and their superior brain capacity, is solely the product of being in the right place right time, is analogous to the thought that the sophisticated genetic coding, that has lead to millions of variations of animal life spanning nearly four billion years, capable of adapting in the face of such global disasters as a massive meteor strike, is the product of some primordial lightning storm striking life into a puddle of random chemicals—To a practical person, that's just too hard to swallow, we are in drastic need of a much better theory.

Accepting the concept that a random set of events, a pool of raw chemicals activated by a lighting storm, resulted in the formation of proteins and DNA which later coalesced into sophisticated multi-cellular, multi-limbed structures, takes an enormous amount of faith. This, I think, is analogous to suggesting that if one gathers together an enormous pile of iron ore, copper ore, limestone, sand, and coarse rock and waits for some really powerful lightning storm to come by, that eventually a hundred story skyscraper, complete with air-conditioning, lighting systems, alarms systems, communications systems, elevators, fire-control systems and computers will rise up out of out of the raw materials all by itself. Honestly, I think I would be waiting an awfully long time. Probably, till the next bang in the oscillating Big Bang theory of the Universe—and its still not going to happen, not with what I am aware of what the nonhuman physical universe that surrounds us is capable of accomplishing. Further, I would venture to say, the human body is a much more sophisticated design project, than required to erect a skyscraper.

Paleontologists and geologists have determined that over the course of the 3.85 billion years that life has existed on the planet, there have been recorded five instances where mass extermination of life has occurred. During the last recorded mass extinction, which occurred just 250 million years ago, it is estimated that ninety percent of the life that existed on the planet at the time, perished. In the face of mass extinction from a cause such as a meteorite striking the planet, a volcano exploding in a cataclysmic eruption, an epic tidal wave caused by the shifting of massive tectonic plates on the ocean floor, or whatever, again it is hard to believe if life were the result of an intricate set of random events, that the random events necessary for life would repeat themselves.

Returning to our pile of iron ore, copper ore, limestone, sand, and rock, still piled on the ground, and I am certain hasn't moved—if a skyscraper were to be erected, the iron would have to somehow become steel, the copper ore refined into copper wire and piping, concrete would have to be formed from the limestone, sand and rock, and then various forms of plastics, glass panels, glues, rubbers, sealants would have to somehow independently form, and form in the right order, building the basement before constructing the roof. To think that if somehow construction were to occur (without the aide of human construction workers), then have the structure suffer, on five separate occasions, catastrophic collapse all the way back down to its foundation, only to rise up a sixth time to become a 100 story skyscraper, is really a stretch of even the most vigorous, overactive imagination.

It would seem much more logical to consider that life on Planet Earth is the result of a very well crafted, master genetic program, introduced almost four billion years ago. From some original DNA, all plant and animal life has evolved. Everything that is considered life, share in the fact, they are carbon based life-forms.

The biologic program introduced 3.85 billion years ago, as a recipe for carbon based life, was written well enough, and had enough variability in the programming, to recognize environmental features and changes to those features to construct life that was capable of adapting to take optimal advantage of those features. Survival in an extremely hazardous environment, has been the obvious mission of the genetic programming.

But the master genetic program was written so well, as to not only cause adaptation of physical features of one level of animal life, but multiple levels of animal and plant life to effect survival.

Higher levels of animal life are the product of complex food chains. For much of the planet's surface, plant life converts the sun's radiant energy into sugars. Animals, referred to as herbivores, consume plant life and convert the sugars and nutrients stored in plants to proteins, the essential building blocks of the body. Higher levels of animal life, carnivores, feed on the herbivores, to consume the proteins generated by the herbivores.

The master biologic program was written so well, as to allow life to survive on the planet, by causing vital adaptations amongst the plant life, the herbivores, and the carnivores, so that the food chains remained intact over time.

The versatility of the master biologic program has been broad enough to have allowed plant and animal ecosystems to develop not just as the product of the radiant energy of the sun, but as the product of sunless thermalgeologic formations. Deep in the oceans, where no sunlight penetrates, the energy radiated by geysers, produces ecosystems similar, but unique from that of the majority of the sun nurtured ecosystems on the planet.

Has there existed then a Master Biologic Program,...a Mother Program,...an Earth Program ('Earth Pro' for short),...a versatile package of complex genetic coding that is responsible for <u>all</u> of the forms of life that has flourished on our planet?

If there is an Earth Program, from where did the genetic code come? Could the code have been delivered by an asteroid that crashed to the Earth? Could a biologic code have been interjected on a barren, virgin Earth 3.85 billion years ago by a space traveler that made a brief stop on our planet...possibly a galaxy traveling Johnny Apple seed? Could life on this planet been the result of a traveler's lunch having been dropped to the ground, or a traveler's cough, or the result of a traveler's footsteps with contaminated shoes,...remember 'One Big Step For Mankind?'.

Could life on this planet been the result of a more organized venture? How about an *Add 'em and Leave* project? Based on the theory that as the speed of space travel approaches the speed of light, the traveler's time clock slows down in relation to time on a fixed position such as the Earth. How much of the stretch of the imagination is it then that a space traveler intentionally encoded the sterile Earth with a biologic program and plans to return to examine the results? Billions of years might pass on Earth, while very little time might pass, possibly only minutes or hours, for the space traveler venturing through space at the speed of light.

Or could the life's genetic code have come from a secret hidden away in our solar system's asteroid belt? Bode's law is an equation that approximates the position of the planets in our solar system. The monumental flaw to Bode's law is the asteroid belt. At the distance from the sun, where the asteroid belt is positioned, a planet is suppose to exist. Could the asteroid belt be the product of some celestial disaster?...A celestial disaster that was colossal enough in magnitude to have resulted in the destruction of the solar system's true fifth planet, billions of years ago. Perhaps life on Earth is the result of work, by the people of the mysterious fifth planet, to salvage life as they knew it, and we are the result of an act to rebuild life on a planet closer to the sun? Should we be scouring the asteroid belt in search of clues to our heritage? Metal fragments from a previous civilization?

A curiosity is that when a human is put into isolation where the external factors are kept constant, and not allowed to 'see' the sun, the human circadian rhythm runs on a twenty-five hour cycle, not a twenty-four hour cycle. The normal human exposed to sunlight, resets their circadian rhythm to continuously adapt to the Earth's twenty-four hour cycle. This more natural twenty-five hour cycle, is it a reflection of the Earth having rotated at a slower rate at some earlier time in its past? Perhaps, it is a reflection of poor design, or a reflection of a sophisticated design that is meant to allow adaptation to the environment. Is it a reflection of the fact humans were designed to live on a planet that rotates on a twenty-five hour cycle, such as our sister planet Mars, which rotates on its axis on a nearly twenty-five hour cycle (24 hours, 37 minutes, and 23 seconds to be exact)? Could Mars have been the original target planet for Earth Program, and something catastrophic occurred in the solar system that resulted in diverting the program to Earth?

Religion and science have battled viciously, for decades, over their positions regarding how life came about on Earth. Why can't we consider for a moment, the possibility that both positions are right. Possibly a superior being designed a complex biologic program that has led to life flourishing in every nook and cranny, to every corner of the planet. The biologic program effectively created life capable of evolving and adapting to the environmental changes that have occurred on the Earth's surface over the last three plus billion years.

Bacteria are single celled organism that support life. Viruses are essentially genetic blueprints stuffed inside a capsule. Viruses are neither alive nor dead, at least by contemporary definition. Viruses invade living animal cells, make their way to the cell's nucleus. Once inside the nucleus, the virus's genetic blueprints incorporate themselves into the living animal cell's DNA. The virus's genetic material then cause the animal cell to divert normal cellular resources to produce multiple copies of the virus. When the viral copies, inside the animal cell, have become too numerous for the animal cell to hold them, the animal cell's wall ruptures. The replicated copies of the virus travel to other cells and repeat the pirateering-like process. (As a side light…If we better understood the instruction coding used by viruses, could we combat the common cold by infecting viruses with STOP code instructions?…Shut off an AIDS infection with packets of STOP code instructions?)

So why couldn't there have existed a **Vironix Particle**? A vironix particle, being a carrier of the blueprints for life, a hybrid of a virus and a bacterial cell. Viruses are unable to replicate on their own. Viruses require a host cell to burrow into and take over cellular control. A bacteria has limited DNA, but has cellular functions and represents life.

Vironix particles could have been interjected into the sterile Earth. Carrying the Master Earth Program, and containing organelles (specialized internal cell structures) that could utilize the raw materials of the virgin Earth terrain, the vironix particles produced life. Instead of producing one form or species of life, like a normal seed, the vironix particles carried the master biologic program, which took advantage of the level of radiant energy from the sun, the air, soil and water temperature, available nutrients, oxygen content of the water and air, and created multiple variations of life that could survive, even thrive in Earth's harsh environment.

Sampling the environmental factors existent on the Earth's surface at any given time, the vironix particles caused life to evolve so as to produce adaptation to the environment to insure the best chance at survival. Survival of the fittest has been a key law of nature. Survival at all costs is probably a more powerful rule.

This same mother program is most likely responsible for the evolution of humans. With such a sophisticated program at work, could the human bloodline be carrying genetic material that plans to evolve into some further design? Is human life so precious because we carry in our genes something that truly is priceless? We are supposedly made in the likeness of God. Maybe we carry God's genetic coding inside ourselves? Is there yet to be a metamorphosis to occur that will explain Mankind's existence?

In a strange twist of fate, are we headed toward an ***Inverse* -Darwin theory**? Will the scientists who bitterly defend Darwin's theory of evolution and have vigorously denounced the existence of a 'god' or a 'creator', someday return to the courtroom and ardently argue to reverse their position?

If we consider that a master biologic program is responsible for life on this planet, then we have to recognize that someone, some party, or something of great intelligence wrote this master biologic program. If science proves that all life on the planet stems from *one* biologic program, then science could *still* argue that a lightning bolt struck a primordial pool of raw chemicals thus initiating evolution; but in effect, science will be actually proving that there is indeed some form of intelligent being or intelligent life in the universe, possibly in our likeness, but superior to our own, that is responsible for creating the life that has inhabited Earth.

As we studied medicine, the mishaps and faults of the human body, i.e. the appendix, cancer, male pattern baldness, congenital deformities, seemed to reflect that humans were designed by some form of committee,…often reverently referred to as the 'celestial design committee' by some of those in the medical profession.

For those who believe it is not possible that a committee might have been responsible for the creation of the Human race, one might ponder Genesis, where it is stated, God said, "Let US create man in OUR likeness, and in OUR own image."

Have you see God lately?

Stop, think for a moment, some may have. If as some religions state, man is created in the likeness of God, and we were to consider, even for a moment, that our life was created by a superior intelligent being or superior beings…then the idea of a master biologic program creating the basis of life generates a physical explanation of how humans exist, perpetuate, and populate the planet without constant physical attention by a superior being or beings.

Well, if we consider for a moment that we are able to formulate visions in our heads, separate from what our eyes actively see (artists and engineers actively do this all the time; those of us who dream—see visual memory files as they dream), then we know there exist memory files that we can retain in the brain and be recalled, to review at desired times. Such recalled visual memory files can be very detailed and can be in color. In my mind, I can change the color of the image of the recalled memory file, and I can change the image shape and position. I can redesign the image and I can store the new image file in my memory to be recalled at a later time. This would seem to suggest an elaborate scratch pad memory which is capable of generating three-dimensional images, most likely located between the occipital and temporal lobes, with inputs from the frontal lobes.

What if the superior being, or superior beings, that generated the master biologic program, inserted memory files that are meant to open up to certain individuals, at certain times, along the evolutionary course of humanity? These master visual memory files could be down-loaded to the occipital lobes from remote memory sites in the brain, and for all intensive purposes, appear to the person who experiences such visual image, just like any real-time image their eyeballs might see. A master visual memory file might be very convincing, may strengthen one's religious convictions, or, even cause a person to change their religion or at least their destiny in life.

No joke,…Maybe some have seen the face of God or at least our Creator or one or more members of a committee of Creators?

As a finale, anybody yet come to the *ultimate conclusion?*

If someone, some Devine Deity, i.e. God, or something, or some committee did write a Master Biologic Computer Program, as far as I can tell, there is only <u>one</u> form of human in existence today, Homo sapiens with 99.9% pure genetic code…therefore, there is only <u>one</u> program…therefore, no matter what religion a group or an individual practices, no matter what name one gives the ultimate being they wish to worship, we are <u>all</u> thanking the <u>same</u> God for the miracle of life.

If the incredible revelation that 'one author' wrote the biologic program that gave life to this planet doesn't bring about world unity and peace,…I'm afraid, possibly nothing will.

Let's all light a candle in our hearts,
   Open, and shine light on the ancient corridors in our minds,
      resolve our superficial differences,
         hope, and *yes*, pray for the gift of understanding,
            which should ultimately lead to peace,
                     For all.

43

YOUR HERITAGE PASSED ON

Computer's can pass on their heritage by downloading files to a disc, which can then be inserted into a newer, faster computer. So, the average freelance writer purchases a computer. Taking advantage of a word processing software, the author writes on many different topics over the course of several years. When technology advances far enough to cause the writer to invest in a newer model computer, the writer simply saves the collection of text files to a disc. Physically inserting the data disc into the new computer, the writer can download the files to the new computer's hard drive. Instead of having to start from scratch with a new machine, the freelance writer can start with the new computer, right where one left off with the old computer.

We have all experienced a sense of deja vu at one time or another. We have all felt that sixth sense of having done something, having performed some task, in exactly the same manner, at sometime in the past. It is a strong sense or feeling when it occurs. It is a feeling that suggests the action could have occurred within the last hour, the last week, the last year, in the last decade. Often it is also a feeling that the action is being repeated repeatedly and that the same action could have occurred centuries before.

The sex organs produce the sex cells. Ovaries in women produce eggs that carry half the genetic material required to create a human. The testicles in a man produce sperm cells. Sperm carry the complementary half of the genetic material required to create a full human. What if the ovaries in a woman and the testicles in a man did more than simply create sex cells? What if the sex organs could also make contributions to the human DNA packaged in the sex cells? What if the most important information learned or most notable experience that one had in their young life, could be translated into the DNA coding, and passed on to future generations through the DNA carried by the sex cells?

The passing on information or data files from one generation to another through the DNA carried by the sex cells may suggest the very importance of caring for and protecting our future generations, which may truly be an extension of ourselves.

44

REVISING THE EDUCATIONAL PROCESS, TO PRODUCE BETTER LEADERS FOR TOMORROW?

The educational processes that I participated in were quite diverse in their philosophies. Elementary school was for the most part rote memory with some individual expression allowed in art class. High school was again for the most part, rote memory with individual expressionalism in art class and literature, and of all places, shop class. Though I was a computer, science and mathematics geek in high school, I enjoyed shop class because it offered a three dimensional medium to express my creativity. My shop teacher didn't mind that I wasn't cut from the regular mold of shop student; thankfully he accepted me in the class and let me have at it in the metal work, woodwork and the foundry. I wasn't totally a geek, I did letter in varsity track and cross-country, but I was awfully close to being a geek; I left high school as a valedictorian—I thrived off of the academic challenges school posed. I really did look forward to shop class, and was truly disappointed when that semester came to a close. Later, following high school, I worked my summers as a plumber's assistant building new homes. The three dimensional skills I had learned in shop class became very useful when I worked building houses.

The undergraduate college educational process was a totally different experience than high school; though, I must thank my high school physics and math teachers in trying to get the students to think beyond rote memory and begin reaching the solutions to problems by thinking 'outside the box'. Being able to think creatively became very useful in my college experience. My undergraduate degree was in electrical engineering. The university I attended was ranked number seven in the country for its engineering program. Prior to attending college, I had thought I would be studying electronics, with classes pertaining to teaching the core elements of what was in a television or radio set. I couldn't have been more wrong.

My earliest classes in electrical engineering were classes regarding theoretical current flow through circuits. The big debate at the time I attended college revolved around the theory: was electric current that ran through wires the result of electrons flowing through the conducting substance, or the flow of holes, the orbital space left by an electrons when an electron left an atom. The classes advanced to the level of mathematical modeling the current flow through circuits. The educational process reached a theoretical summit in my fourth year in a required class simply called *Communications*. In this class, we studied the mathematics of Laplaise and Fourier transforms and how they were used to generate modulations of carrier waves to produce television signal waves. Five graduate students attended the class, the rest were undergraduate students, like myself. The overwhelming sense amongst the undergraduates was that we were, for the most part, lost in the mathematical theory, despite having had a solid foundation provided by three intense years of mathematics in upper level calculus and beyond.

When I received a 'C' grade in the class, which meant I had completed the class successfully and could therefore graduate as an electrical engineer, I found my professor and thanked him for the grade. He opened up his record book and showed me the grades of my fellow students, many of which had not done as well as I had in the class. I recall feeling so lost, yet so relieved.

My electrical engineering experience was far from rote memory. Oftentimes, our professors conducted open book tests. We were allowed to bring any textbook resource we wished into an open book test. The idea of engineering tests was often not to test the students on what one could memorize, but could the student successfully apply what was learned to solve new problems, problems the students hand not yet encountered in class.

I remember one test in particular, in my third year, when I had sixty minutes to complete a four question electrical engineering exam. Each of the four questions was a story problem and each problem required a written answer or mathematical equation in order to gain credit for answering the problem. To answer one question correctly meant a 25% grade, two questions correctly meant 50%, three questions 75% and of course to answer all four correctly meant a 100% grade for the test. Upon receiving my copy of the test at the beginning of the hour, I carefully read the first question. I quickly realized I didn't know how to answer the question, so I went on to the second question. After reading the second question I nervously realized I did not know how to answer the second question either. As the minutes ticked off the clock mounted at the head of the classroom, I carefully read the third question. I painfully learned I could not solve the third question either. I scratched out a few mathematical equations trying to lead myself to a solution to the third problem, but I knew my approach was incorrect. In desperation, I read the fourth question, figuring that if only I could get the fourth question right I could at least get 25% on my test. I struggled at solving the fourth question for the next thirty minutes and was not able to arrive at a solution to the fourth problem. I looked up at the clock on the wall and realized I had only eight minutes left.

I had yet to arrive at any form of an answer for any of the four questions. It was an open book test, and three textbooks sat on the floor next to me. Like most of the students in the room taking the exam, the texts were never opened. The answers to the four questions we were challenged to solve for this final exam were not to be found in our books. These four questions were applicable to our class, but were unlike any problems we had previously worked through for this particular class. It was the engineering way.

I remember staring down at a nearly blank test paper with less than eight minutes to go. I envisioned having to take the class over again due to unarguably flunking the final exam. I flipped the paper over and reread the first question. At that moment my subconscious kicked into what seemed like overdrive. The answer to the first question flooded into my conscious brain. Quickly I scratched out the appropriate text to answer the first story problem. When I was done, without looking up at the clock I reread the second question. Bam, the mathematical equation that was required to answer the second question flooded into my conscious brain and I hurriedly scratched it out on the test paper. As I flipped the paper over I glanced up at the clock on the wall. Four minutes remained. I quickly reread the third test question and like an explosion, the answer to the third question rushed into my conscious brain. My hand worked feverishly to write out the answer on the sheet of paper. With two minutes remaining, I reread the fourth question. The fourth and final test question required a step by step process to be written out in a series of mathematical equations to arrive at and prove the answer to the problem. I had struggled for thirty minutes trying to come up with even simply an approach to the problem, much less the answer to problem number four, and had been unsuccessful in my first attempts. Now with two minutes left, my palms sweaty and my heart racing, pumping at one hundred and thirty beats per minute, I glared down at the words that comprised the last test question. With sixty seconds left, my subconscious flooded my conscious brain with an answer to the question. I scribbled with my pencil like my hand was on fire. I quickly wrote down mathematical equations in the steps to properly solve the engineering problem. I was two thirds of the way done when the clock reached the sixty minute mark. The instructor called for a halt to the test. I placed my pencil down, knowing I was only seconds away from scrolling out a complete answer to the fourth test question.

As the test papers were collected by the instructor, I remember feeling a sense of emptiness for not having fully completed the fourth test question, but also, at the same time, a sense of awe for the power of the human brain, realizing that I had gone from a failing test grade of 'zero' to at least a 'passing' grade inside the last eight minutes of the sixty minute test. My final score on the test was an eighty-seven out of a possible hundred points.

The medical school experience couldn't have been more divergent from the engineering experience. Medical school was nearly purely rote memory. The medical school experience involved learning volumes upon volumes of medical facts. The testing was for the most part consisted of written tests comprised of multiple choice questions. The average test was conducted over a four hour long period, which were usually conducted on Saturday mornings. The possible answers were multiple choice, but most of the tests were comprised of k questions. K questions were designed to challenge the test taker to choose from the first answer, second answer, third answer, or a combination of the first and third answers, or to choose from all of the answers, or finally, to chose that none of the first three answers were correct. Memorization of the facts and sequences of facts was very important to successfully completing medical school and subsequent residency and fellowship training.

Where the engineering experience was designed to produce imaginative and creative individuals who were trained to solve new problems, the medical school experience was designed to produce individuals that acquired and amassed an extensive library of knowledge that could be accessed, instantly by the student, on an as needed basis.

We were able to challenge the answers to questions on our medical tests. Memorization was so much the focus that imaginative thinking was not of particular value. On one exam, I challenged the correct answer to a question regarding the effects of a pharmacologic agent called Antabuse, a drug used to deter alcoholics from drinking, suggesting that my answer, consider an incorrect choice, was in fact an alternative correct answer. I located a chemical formula in a dusty pharmacology text I found in the medical school library that supported my position regarding the answer. Despite stapling a photocopy of the formula copied directly from the text, and writing a lengthy argument, the challenge I had submitted was not accepted by the faculty member who administered the test.

The above-mentioned concepts are meant to illustrate the two diverse means of education *memorization of practical facts* versus *abstract problem solving in a mathematical medium.*

Consider for a moment, that the basic concepts set forth in this book are valid. The future education process and future job market may involve attempting to actively identify individuals that have good random access memories (efficient memorizers), versus individuals that arrange information in tabular form, versus individuals that feature imagination generation units, versus individuals that are able to program their own intelligent agents, versus individuals that have the closet access to the genius files. We should also be encouraging individuals to master more than one major educational discipline. In the past we have suggested that those that studied a profession and studied the arts were more rounded individuals. Actually, what was probably being recognized was that the broader the information base to which an individual was exposed, the more data files with which the imagination generation unit had to work. The closer an individual was to an appreciation of music, the more stimulation they received, the more chance that unlocked genius files would open and be brought forth from archived memory to the awareness of the conscious mind.

Alas, amassing large quantities of knowledge is only as good as one's capacity and desire to understand the knowledge, and apply this precious gift toward solving contemporary problems.

45

OPENING NEW FRONTIERS IN COMPUTERS AND ROBOTICS

At the heart of present day computers are silicon chips comprised of millions upon millions of transistors, a sophisticated term for on/off switch. The advances over the last decade have been to compact more and more transistors into tighter and tighter spaces. The advancement has been one-dimensional.

Understanding both the microscopic construction of nerves and the macroconstruction of nerves, nerve clusters and nerve nuclei will help us further our development of our computer technology in multidimensional planes. Studying the design of the human brain and how it interacts with the remainder of the body, will further both medical science and robotics, both in theory and design.

The human brain offers secrets of construction and design that evolution has perfected over the course of billions of years. Studying and unlocking the mysteries of the human brain such as the software language and organization of the Brain Operating System Software (BOSS), will lead us to newer and more challenging frontiers. The broader the education of the members of our research teams, the better equipped our newer generation of computer and robotics designers will be to meet the challenges of creating newer, better, faster, smaller technologic designs.

Cross-pollinating our knowledge between engineering, computer science and medicine can lead to a magnitude of untold advances in computer science, computer architecture, and robotic designs that may reach well beyond our wildest technologic dreams.

46

NEUROCOMPUTEROLOGY IN SEARCH OF EARTH PRO

Understanding how the pieces of the brain function, how the brain retains memories, transfers information, how the brain's logic units assimilate and analyze data are all important frontiers to explore. However, cracking the brain operating system software code and deciphering the computer language and commands the brain uses to effect its functions, would be the ultimate feat. To learn the software code would open up incredible access to the brain.

Medicine is filled with terms that are comprised of splicing root words together into large, difficult to pronounce, grossly expanded descriptive terms. The term neurocomputerology is comprised of the root *neuro* meaning brain and nervous system; *computer* referring obviously to computers; and—*ology* meaning 'the study of'. So the term, neurocomputerology (NCO), refers to the study of how the human brain and nervous system functions like a desktop computer. Neurocomputerology would also apply to the research efforts toward developing utilities to communicate with the human brain and decipher normal and abnormal brain operating system software. Neurocomputerology could also lead to developing means for humans to communicate with the brains of other forms of life that share the planet with us.

Search For Earth Program

The question remains,…Where do we start?

The pages of this text have led one down the road of pondering the possibility that the human brain is a sophisticated biologic computer.

But, from a practical sense, where do we go from there?

It was mention at the onset of describing the human brain, that peering into the brain is a much more difficult task, than studying the inner workings of a computer. You can't just take a screwdriver, crack open the brain, have a look around, and not expect the brain to suffer major trauma.

In addition, there is no keyboard or mouse or video screen with which to make contact with the brain operating software system. There might be the capacity to directly link to brain tissue through a form of sensor probe or data interface device. Such a means would require an invasive procedure. Such invasiveness would create a risk of infection and/or brain damage.

One might think in terms of communicating with the brain operating system software through an infrared-like data transfer device. Data, commands or programs, might be up-loaded into the brain through the eyes using light signals. The occipital lobes or other portion of the brain might be able to interpret such light impulses. The eyes do not emit an output signal that we know of, so an alternative means of reading data output from the brain would be required. Pulsed sound waves might be an alternative means of accessing the brain operating software. Output from the brain operating system, that is responses to the light or sound inputs, might be by means of monitoring the electrical activity of the brain (through a device such as an EEG, short for electroencephalogram), by monitoring muscle reactions such as blinking or limb movements, or by recording sounds generated by the vocal cords.

But how would we even start to try to communicate with the brain's operating system? How would we be able to accurately recognize what a response might be?…Much less be able to interpret a response.

Remember the fascinating movie CONTACT written by Carl Sagan, directed by Robert Zemeckis, featuring Jodie Foster? The story that the movie CONTACT told, involved scientists searching the heavens with radio telescopes. These scientists, lead by Jodie Foster, record a message from outer space. The message is comprised of a code. Once deciphered and pieced together properly, the code provides blueprints for constructing an intergalactic transport device which, at the end of the movie, sends Jodie Foster across the galaxy. Very creative storyline, a movie one should see, if they have not already done so.

But thinking about it, do we have to wait until we receive a message from the dark recesses of outer space?…or are we carrying Earth Program around with us?

A giant red wood pine is created from the sprouting of a single seed. The thick trunk, the expansive branches, thousands of leaves, years of survival in a harsh world, are all the product of the genetics of one seed.

If one seed can create a colossal red wood tree, is it that much of the stretch of the imagination that one biologic program was responsible for the creation of life on the planet? Marine life might be one branch, insect life might be another branch, birds would represent a third branch, cold blooded animals a fourth branch, mammals a fifth branch of a colossal tree that sprouted from one biologic program.

Fact: (1) greater than one billion base pairs of the human genome, one third of the human genome, have been mapped out. The remainder of the genome is actively being defined.

Fact:(2) Life on planet Earth share genes. As mentioned earlier, humans share some genes with a banana. This makes sense. Once basic designs such as creating cellular structure were written into the genetic code, such vital designs were shared amongst the forms of life, instead of being continuously recreated in other forms.

Fact:(3) As mentioned previously, HOX genes are master control genes are shared by life on the planet, that turn 'on' and turn 'off' physical characteristics, creating wings, legs, or fins depending upon the environment in which life is required to survive.

If one entertains the thought that a single master program has been responsible for life on planet Earth, then what would one get if we combined the information gathered from the various life forms that exist on the planet? That is, if we took the human genome, once it has been fully deciphered, and overlap the information with the genetic code of all of the other life forms that exist on the planet, would the result be Earth Program?

We could take this mega-conglomerate of biologic code, and remove redundant instructions and data. Then the genetic code could be broken down into definable pieces, such as the genetic code for red hair, the genetic code for floppy ears, the genetic code for blue eyes, the genetic code for fins, legs, wings, etc. We would be able to create a library of physical characteristics associated with genes. Using this library of genes associated with physical characteristics, we could formulate an understanding of the meaning of how the genetic code was written and possibly be able to predict the meaning of the genetic code. A mysterious, lingering question, is: Is there extra code stored with the genetic code necessary to create life on the planet? What would the meaning of the presence of extra code be? Extra code could provide a template to help us decipher the existing code, or could represent the storage of data that humans will need to use at some point in the future, or it could represent the signature of the creator himself or herself.

If one returns to the world of the desktop computer, as 'I' type on the keyboard to produce this text, the computer is using ASCII code in order to decipher my keystrokes, to eventually create the correct letters for 'I' and 'you' to view. The ASCII stands for American Standard Code for Information Interchange. Using ASCII standards, if 'I' type a capital 'A' into the keyboard, this has the number '65' designated to it, which, in its binary equivalent at the level of machine language would be '1000001'.

Let's speculate for a moment, that no one on Earth had ever seen a computer, and a computer happened to fall out of the sky into the hands of the scientific community. If humans were able to extract information from the memory of the computer, and all that humans were able to obtain were the machine code of ones and zeros, it would be very difficult to decipher the meaning of the ones and zeros. If investigators could assign meaning to portions or defined strings of ones and zeros, possibly the meaning of the code could be worked out. That is, if one who had never seen a computer before, could decipher that repetitions of the string '100001' meant '65' or better, represented a capital 'A', then one could begin to put larger sets of information together, eventually stringing letters together to make words, and then sentences from words (obviously grammar, or understanding the rules of how the words (commands) are to be strung together, to create proper meaning would be important-this is why there might be extra code in the DNA, in order to provide documented rules of grammar).

Human 'genes' are coded in base four, not binary code. The challenge to break the meaning of human genetic code is a daunting task to say the least.

But if the software code used by the human DNA can be deciphered, it may lead to being able to decipher and read the software code used by the brain operating system. Once we have some basic knowledge of the software code, we might be able to attempt to communicate directly with the brain operating system. It is conceivable that animal life on the planet share a similar brain operating system. Lower forms of animal life, may simply lack the processing power and memory capacity that humans have been afforded. The capability to communicate with the brain's operating system, directly, would be an incredible step forward for human medicine, veterinary medicine and technology.

Those brilliant and creative men and women that would pursue this fascinating and exhilarating offshoot of medical-technical science, might be known as 'Neurocomputerologists'.

Neurocomputerology could be the most exciting and rewarding unexplored frontier humans have yet to encounter, and we carry it with us, wherever we go.

47

THE HUMAN COMPUTER—GET THE MOST OUT OF YOURS!

The human computer has a wide variety of peripheral devices it is required to interface with, on a day-to-day basis, as well as a minute-by-minute basis. These peripheral devices include biologic devices that exist inside and outside the brain.

Inside the brain includes the conscious, subconscious and memory units. The subconscious Human computer processor interacts and works in partnership with the frontal lobes (consciousness) of the brain. Our consciousness works on all of the higher level thinking that is required to make us individuals, while the subconscious toils with all of the lower level thought processes that are required to make the body run properly.

The subconscious is the go between the consciousness and our memory units. The subconscious also interacts with the emotional unit in the brain, sexual drive, the logic unit, and the peripheral devices located inside and outside the brain.

The peripheral devices outside the brain include internal and external sensory devices, sound generation by the vocal cords, muscle function, and other bodily functions such as devices to regulate body temperature, body blood sugar, blood pressure, body water balance, energy conversion. The external sensory devices include optics, audio perception, sense of smell, sense of taste, and a sense of touch, position and pain.

376

The memory unit is divided into four major components. These four components include Real-time Memory, Scratch Pad Memory, Remote Personal Memory, and Archived Memory. The Real-time memory is responsible for holding vast amounts of personal data that we need to recall quickly such as our name, age, time of day, address, telephone number, where we work and many other bits of information. Scratch Pad Memory is a location in our brain for storage of short-term facts or numbers. Scratch Pad Memory is often used in conjunction with the Logic unit. Scratch Pad Memory is used when we are adding two numbers together in a mathematical calculation. It is where we might store a telephone number after we find the number in the telephone book and then dial the number. Scratch Pad Memory clears the facts and numbers stored in it on an ongoing basis. When requested, the data stored in the Scratch Pad Memory may be transferred to Real-time Memory or even to the Remote Personal Memory. The function of the Scratch Pad Memory is to take facts, rearrange them, work on them in order to come up with solutions to mathematical or practical problems.

Remote Personal Memory is a library of experiences, names and faces of individuals we have met, facts that we learn in school or career training. This memory may take time to recall the data. It may not be readily accessible at a moments notice. So the conscious brain may request the subconscious computer to ask the Remote Personal Memory unit to recall the street address of a house that one lived in as a child. A fact that could be twenty-five years old if the individual is thirty-nine. This old address may not be readily available when the conscious mind asks for it. The data regarding this old address may have been moved from the Real-time Memory, because this address is no longer in use, to the Remote Personal Memory unit. When the subconscious asks the RPM unit to retrieve the old address, the RPM may have to hunt through the vast files of personal memories to find the requested bit of data. So an hour after the request is made to the RPM unit, this unit may find the data file regarding the old address, send the data file to the subconscious, which then relays the information to the conscious mind. To us, suddenly the old address flashes in our head.

The Archived Memory unit is quite a different and unique set of memory files. This part of our memory represents files that are not our own. The Archived Memory unit is broken into two subdivisions that may on the surface appear to be totally separate, but it actually may provide similar function to the human species. One subdivision is Instinctual Memories. These are memory files of a rudimentary nature. The instinctual files are available to an individual when learned human behavior is not available or social rules and regulations are not available. These files include basic information on feeding oneself, avoiding danger, being aggressive to protect oneself, basic sexual interaction and function, and caring for offspring. These rudimentary files are not necessarily very detailed, but provide essential information for a primitive human being to survive in the wild amongst animal predators and successfully reproduce. These files would have been essential to the survival of man hundreds of thousands of years ago, prior to the dawning of civilization, which has only occurred with in the last ten thousand years.

The second component of the Archived Memory may be what humans refer to as imagination. That is, all humans contain a portion of their brain or memory that they do not actively use. It is often referred to as the right side of the brain that is not used (if one is right-handed). This portion of the brain that is not actively used may represent a vast library of facts and figures preset in our brains. This part of our brains would be referred to as the Genius Files. That imagination is the result of (1) a random fact generator and/or (2) the subconscious brain accessing a data file stored in the Genius File library located in the Archived Memory portion of our brain. A portion of the inventions mankind has devised over the centuries may have been derived from data files already stored in our brains.

This scenario would explain why there are Geniuses in the world. True Geniuses may be simply individuals that have better or freer access to their Imagination Generation Unit and/or Archived Memory files, and actively take advantage of their outputs. In addition, brilliant figures in man's history such as Hippocrates, Leonardo De Vinci, Michelangelo, Copernicus, Newton, Einstein may have been individuals that trained their subconscious minds to repeatedly and actively access the Archived Memory files. These futuristic thinkers of our past may simply been better in touch with the Genius Files then most of their contemporaries.

Being able to understand that we have capacities in ourselves is the first step toward learning how to best utilize them. As the future unfolds learning what mental tools we possess may be critical to one's success and fortune. Learn what you can do with your human computer, to train and explore the recesses of your mind—so as to:

—*Get the Most out of Yours.*

Best of luck.

EPILOGUE

It is not the intention of the author to sound so naive as to suggest or even appear to suggest that a glorified circuit board, no matter how well programmed, could ever replace a human,…any human.

Men and women possess unique features that set us far apart from the wildest dreams of the sci-fi techno-mechanized world.

Humans possess sympathy and empathy, we possess the capacity to appreciate our surroundings, each other, and the gift of life. We have attitude, we drive ourselves well beyond manufacturer's specifications, in our passion that no obstacle is too great for us to conquer. We have spirit, we can focus our energy and strength to achieve truly great things. We exhibit responsibility for each other, we strive to grow and become strong, while trying to look out for our brothers and sisters. We have each other and we know this. We gain strength in numbers, we are human and no machine could ever take our place in the universe.

Unfortunately, medical science has not yet progressed far enough to establish that the contents of this book are factual beyond the description of the computer architecture and the description of the anatomy of the human brain. Most of what is written in this text regarding how the human mind functions is purely speculative. By observing humans and how they function and respond to their environment, and given our lack of concrete science, speculation and drawing analogies to elements in our surrounding environment are the best tool we have for analyzing ourselves, given the current state of technology and medicine.

Speculation usually occurs before science has a chance to prove or disprove theories. Early theories are generally referred to as a hypothesis. This book, the Human Computer: Get the Most Out of Yours, is a hypothesis of who we are, how our brain works and…hopefully how we can get the most out of the fabric and framework of our minds. As time advances and medicine progresses, speculation hopefully will evolve into reality.

Aptitude testing by both the school systems and employers should possibly be modified. For corporations wishing to hire employees with certain job performance skills, current testing methods may be adequate. But for corporate leaders searching for individuals that possess the capacity to access memory data files that have yet to be unlocked by others in the population, they should be testing for potential geniuses not just in regards to their skill performance, but in an individual's capacity to search their memory files for new and innovative ideas. With the proper encouragement individuals that can access remote memory files could be very valuable to society. Training future generations to access remote memory files versus having them assume new ideas are simply the result of a series of random events that happen to befall only a chosen few of us, could lead to solving some of society's most critical problems. Being able to access remote memory files at fortuitous times could represent untold riches for the individual who knows how to successfully develop the information.

The underlying current flowing through the pages of this book begs to ask the question, 'What untold secrets do we have locked away in the data files of our brains that have yet to be discovered or even dreamed about?—And is there a clock ticking away, that challenges the essence of human survival, that demands that we actively explore the riches hidden away in each and everyone's brain? Open your mind, know who you are, realize what secrets you possess,…Be the next genus or world leader or praised entrepreneur…The world is expecting you.

Finally, hopefully this book has sufficiently demonstrated that the body is the product of an elaborate computer program that creates the multi-limbed, multi-organ, human infant from two sex cells....

   -during a human's maturation, biologic computer programs are at work collecting vast quantities of information and overseeing development of the body's internal and external structures...

   -sophisticated computer programs defend the body from intruders...

      -in our middle-age, feedback loop software equations maintain our bodies in working order....

         -programs drive us in our choice of a mate, career, and that which we actively seek for pleasure....

            -and in our older age, programs fail...leading to programmed cell death....

               -the more we know about our computer, the better we can be at utilizing and maintaining such a wonderful system.

As a last thought,.....

   the *oldest* known profession, is <u>not</u> prostitution,.....

                  but a *computer programmer*,
                  smile and take pride.

Ω
*THE PERFECT
ENDING TO THE
PERFECT CRIME*

Dressed smartly in a custom made silk suit, David Thomas confidently marched into the boardroom. David Thomas, a forty-eight year old, international entrepreneur, president of a high-tech microchip company stepped up to the front of the room. He was expecting a boardroom filled with potential customers. He had hopes of securing a sales deal worth one hundred and ten million dollars.

David Thomas's mahogany eyes scanned the ten men that stood in the room. The men surrounded a large oak oval table that stood in the center of the conference room. His nearly hundred years of experience gave him a keen sense of deciphering body language. The men that stood in the room were dressed as if they were business executives, but David Thomas's sixth sense keyed him in on the fact that a hidden agenda was about to be uncovered.

A tall, sandy haired man broke the ranks of the rigid formation and stepped forward. In an English accent, he asked, "Are you Nevin Thomas?"

David Thomas stood tall and confidently replied, "No, I am David Thomas. Nevin was my uncle. But he died thirty years ago."

The sandy haired man produced a wallet, flipped it open revealing a shining metal star, and said, "Mr. David Thomas, or I should say Mr. Nevin Thomas…I am Rick Temple,…from 'I squared PS'…You are under arrest."

At the announcement by agent Temple, the others present in the room took their cue. Guns were retrieved from concealment. The men moved to block any attempt of escape by their target and they quickly made it known that they were prepared to neutralize any attempt the fugitive might create to turn the arrest into a violent confrontation.

Taken by surprise, David Thomas threw up his arms. He was quickly subdued by the team of agents that were more than ready to take the fugitive into custody.

As his wrists were being shackled behind his back, David Thomas stood defiant. In a deep, unwavering voice, he asked, "What is the meaning of all this?" David Thomas leaned over and peered at agent Temple's identification and angrily asked, "And who are you guys?"

In a stiff voice, agent Temple said, "I said 'I <u>squared</u> PS'," he added, "International Identity Police Service."

David Thomas asked, "What do you want with me?"

In his English accent agent Temple responded, "You are believed to have stolen the identity and the body of David Thomas."

"I've done no such thing."

"Quite the contraire, we believe you are none other than Nevin Thomas in David Thomas's body." Agent Temple watched closely, monitoring his suspect for any change in body language that would help confirm the acquisition. A micro-camera and microphone hidden in a pin attached to the agent's lapel recorded the entire encounter.

David Thomas nearly choked. His mouth suddenly grew dry. His pupils widened, his eyebrows rose to the top of his head.

The I^2 PS agent asked, "Any comment, Nevin Thomas?"

Caught completely off guard, Nevi Thomas stammered, "You must,…I mean you are mistaken."

Agent Temple responded, "We will not be mistaken once we've had the chance to scan your brain and study your BOSS's deep rooted primary memory files. No matter how much you have come to believe that you are David Thomas, the underlying software files that established your protocols are still intact."

"That's absurd…It can't be accurate."

"On the contrary Nevin,…We have the capacity to define the memory protocols that your brain software used nearly one hundred years ago and compare them to protocols that David's brain should have developed. I am certain we will find some of David's original protocols, still intact in remote memory files that we can use to specifically prove that your brain has two operating systems,…which is physically impossible,…unless *you* super-imposed your consciousness into your nephew's brain."

Nevin Thomas said, "That's not possible."

Agent Temple retorted, "Well, what's certainly not impossible is getting a CAT scan of your brain and searching for plug holes in your skull."

Reflexively, Nevin Thomas tried to reach up and touch the back of his scalp where an old scar was etched across the skin, but the shackles prevented it. Agent Temple took notice of the subconscious gesture.

Nevin realized what he had done. He quickly dropped his hands back down. The chain connecting the handcuffs clicked as the metal links struck each other.

Agent Temple stared at Nevin, as he stated with confidence, "We already have the court order."

Nevin blurted out, "You'll never…" then the man's voice trailed off. He bit his lip as he tried not to implicate himself.

The I^2 PS agent knew what Nevin was going to say. He responded, "Don't have to stick a physical plug into your head." He was quick to add, "We now use Flash Cortical Optics technology."

A bewildered look appeared on Nevin's face.

Agent Temple filled him in, "F.C.O. technology allows us to use light signals flashed through the eyes to establish contact with a brain's operating system. We can download information by detecting electrical impulses picked up by a modified EEG brain wave monitoring device adjusted to detect oscillating alpha waves."

As the agents led David Thomas's body out of the boardroom Nevin Thomas thought about the last few statements the agent had made, and asked, "What if your fancy scanners do what you say they can…Ah…What would such an identity thief expect?"

As they marched down the hallway, Agent Temple answered, "Justice would dictate that we remove their consciousness from the victim's body."

"And do what with the consciousness?"

"Typically cases such as this, the criminal's consciousness is downloaded to an artificial memory storage device."

"For how long?" asked Nevin.

"A capacitor is attached to the memory device, charged to the criminal's expected remaining lifetime."

"And?"

"The capacitor discharges over time."

"And once the capacitor has discharged?"

"The memory device loses power, the criminal's consciousness is lost."

Abruptly Nevin blurted out, "And how humane is that?"

"Actually it's becoming the incarceration method of choice. No elaborate fences, no security risks, little chance of escape by the prisoner...You can store a whole lot of prisoners in a very small space,...And oh,...the cost is right," Agent Temple was quick to add, "The whole process, including the extinguishing phase is quite humane because it is quite painless."

Changing tracts, Nevin asked, "And what would happen to the victim's body?"

"Once the brain is cleared, the body would be donated,...So that a victim of severe third degree burns or victim of some other catastrophe could have their consciousness downloaded into the body to give the person a second chance at a real life."

As the team of agents led David Thomas from the building to the cars waiting outside, Nevin realized that this was an elaborate effort on the part of the authorities. He sensed there was very little chance that he was going to get out of this. Nevin asked, "How did you find me?"

As agent Temple ushered David Thomas toward one of the black limousines, he responded, "Someone in the life insurance company that you ripped off thirty years ago, reopened your case." He added, "They found remnants of a scalp plug amongst the fire debris locked away in storage."

Nevin blurted out, "How would they have every recovered that after all this time?"

"Locked away inside an evidence box somewhere...You know those boys, they keep everything...for just about forever." Agent Temple paused as he opened the rear door of one of the black limousines, then added, "Once the plug was found, your body Nevin, was exhumed."

"Exhumed?"

"They dug your body out of the ground." He quickly added, "They found evidence of previous brain surgery…Holes in the skull where computer access plugs could have been inserted, that the coroner thirty years ago didn't think of looking for."

In a disgusted voice Nevin stammered, "Why would anyone go through such trouble over a thirty year old case that's been forgotten about?"

Agent Temple gently placed the palm of his hand on David Thomas's head and assisted him down into the police vehicle. He smiled as he answered, "Money my friend…Money." He added, "There is fifty million dollars at stake,…plus interest." He commented, "You took those insurance boys for a whole lot of dough Nevin…That's something they never forget…Even after thirty years." Agent Temple added, "And your company's assets look very attractive to them."

A streak of anger flashed across Nevin Thomas's face.

Agent Temple saw to it that his prisoner was properly secured in the police vehicle.

Staring Nevin Thomas in the eyes, agent Temple inquired, "So Nevin, what did you do with David's consciousness,…Download him to some computer hard drive?"

A long cold moment dragged on…what seemed like eternity passed. The skies above were a fluffy dark gray. A misty rain fell down onto the I^2 PS agent.

Finally, thirty years of guilt boiled up from his subconscious and cracked a geyser of steam through Nevin's rigid exterior emotional casing. Nevin, in a disparaging manner muttered, "Naw, he was just erased."

That's all the special agent needed to hear. Agent Temple shut the car door. A sense of finality resonated as the door lock snapped into the latch. With the pat of his hand on the roof of the car, the agent signaled the driver. The police vehicle drove off.

Agent Temple straightened up. He reached inside his trench coat and briefly hit the rewind button. A moment of tape zipped by, then releasing the button Nevin Thomas's taped voice played again, "Naw, he was just erased."

Satisfied, agent Temple turned the recorder off. A gentle rain fell onto his face from the gray sky above. For a brief moment, the lawman sensed they were tears of joy.

One of the uniform police officers that had accompanied Agent Temple on the arrest, strolled up to him. Trying to strike up a conversation the officer remarked, "Do you think the case is in the bag?"

Agent Temple turned and bluntly said, "We got our man."

"Will it hold up in court?" The uniform cop stumbled with his words as he continued, "Get a,...I mean,...Do you think you'll get a conviction?"

With confidence, Agent Temple remarked, "We'll take Mr. Thomas down to the station," he paused briefly, then added, "Tap into his brain...Unzip the memory files in the frontal and temporal lobes, download the visual files from the occipital lobes. Soon we'll know every detail of the crime...That he knows."

Naively, the uniform cop asked, "Then, what's with the tape recorder?"

The special agent smiled. As the rain drizzled down from above he said, "The high tech brain interface-analysis software tells us everything we want to know,...but takes the fun out of solving crimes." Agent Thomas paused, then added, "Sometimes it's still nice to get a confession the old fashion way."

Agent Thomas shoved the hand that held the tape recorder into the side pocket of his trench coat to protect the recorder from the rain.

Politely the officer asked, "Can I give you a lift to the station?"

The special agent tipped the wet brim of his hat and said, "No thanks,...It's only a couple blocks, I'll walk." Agent Temple turned and strolled off down the to sidewalk in the direction of the police station, his figure eventually disappearing into the mist created by the fine rain falling from the fluffy gray sky that lazily hovered over the city.

THE GLOSSARY OF MEDICAL AND TECHNICAL TERMINOLOGY

amp the abbreviated form of the term ampere.

ampere the basic unit of current flow in the practical (mksa) system of units. One ampere of current corresponds to the passage of 6.24 billion billion electrons in one second.

amplifier a device for increasing the power associated with a phenomenon without appreciably altering its quality.

amplifier, audio frequency, any device designed to amplify signals consisting of frequency components within the range of 20 cycles to 20,000 cycles, such as voice or music.

amygdala collection of nuclei which forms an important behavior and emotion center of the limbic system.

anode the positive pole or electrode of a device. In a vacuum tube, it is the positive plate to which the principal electron stream flows.

antenna a physical device for the radiation or reception of radio waves.

aphasia refers to partial or total loss of the ability to articulate ideas or comprehend spoken or written language.

arithmetic logic unit (ALU) the central part of a microprocessor that manipulates the data the microprocessor receives.

ascending reticular activating system (ARAS) components of the brainstem reticular formation that project to parts of the thalamus and subthalamus and pace the activity of the cerebral cortex which if interrupted at the midbrain causes coma to result. Sleep centers in the pons, medulla and hypothalamus project to the ARAS to turn it off to induce sleep.

axon nerve cell process conducting impulses away from the cell body.

ballismus violent jerking or flinging movements of proximal parts of the limbs and shoulders and pelvic girdle musculature, associated with lesions of the subthalamic nucleus.

basal ganglia refers to the large, strongly interconnected nuclear masses of the corpus striatum (located in the cerebral hemisphere), the subthalamic nucleus (in the diencephalon) and the substantia niagra (in the midbrain). The corpus striatum consists of the caudate nucleus and the lentiform nucleus. The lentiform nucleus consists of the putamen and the globus pallidus.

binary consists of the two integers 'zero' and 'one'; is the basis of operation for computers.

Broca area opercular and triangular parts of the inferior frontal gyrus in dominant hemisphere associated with the creation of speech with motor programs for the pronunciation of words and creation of grammar.

buffer refers to a memory device that holds information. Information can be downloaded from a high-speed device, such as a CPU, to this intermediary memory device, until a slower operating device, such as a printer, has the opportunity to utilize the information, thus freeing up the high speed device.

bug 1. A flaw in software that causes a program to malfunction. 2. In medicine it often refers to the flu or viral syndrome.

bus 1. the circuitry and chips that manage the transfer of data from one device to another. A data bus often refers to the cable connection that transmits data information from one device to another. 2. A big yellow vehicle that carries children to and from school. 3. May be considered a pathway in the brain that transfers information from one part of the brain to another part of the brain.

cache a block of high-speed memory where data is copied when it is retrieved from RAM. A cache acts as a buffer, tending to hold frequently used data that would otherwise have to be retrieved from a memory source operating at a slower speed.

capacitor an electronic device consisting of two insulated conductors that is capable of storing an electric charge.

cathode general term for a negative electrode, in an electron tube the electrode through which the primary stream of electrons is emitted and enters the inter-electrode space.

cathode grid a suppressor grid located between the cathode and anode in an electron tube.

cathode ray 1. a stream of electrons emitted by the cathode of a gas discharge tube when the cathode is bombarded by a stream of positive ions. 2. Any stream of electrons as in a tube in which the source of the electrons is heated filament.

cathode-ray current a current in a vacuum or in a rarefied gas formed by the movement of negatively charged electrons.

cerebral ischemia decreased blood supply to the brain.

cerebrospinal fluid clear, colorless liquid secreted by the choroid plexus and in the ventricular system and subarachnoid space; the total volume is approximately 150 ml; rate of formation is approximately 500 ml a day.

cingulum a large association bundle padding longitudinally in the white matter of the brain in the cingulate gyrus that connects the frontal, parietal and occipital lobes with the parahippocampal gyrus and the adjacent temporal cortex.

circadian rhythm biologic activity that occurs on a 25 hour cycle, that is continuously reset by sunlight to a 24 hour cycle. The biologic clock resides in the suprachiasmatic nucleus of the hypothalamus.

circuit in electronics an electrical network providing one or more closed paths for electron flow. In the human brain it is considered a pathway of neurotransmission.

CNS abbreviated form of central nervous system, of, or pertaining to the brain and its neurons.

cochlea a spiral structure located in the inner ear that interprets sound. The cochlea might be considered an auditory analog to digital converter. The output of the cochlea is the eight cranial nerve that traverses to the brainstem.

coma in electronics an aberration of spherical lenses occurring in the case of oblique incidence, when the bundle of rays forming the image is unsymmetrical. The image of a point is comet-shaped, hence the name. Coma regarding the human brain denotes a state of unresponsiveness.

command in electronic computers is (a) one of a set of several signals which occurs as the result of an instruction; the commands initiate the individual steps which form the process of executing the instructions.

(b) an independent signal from which the dependent signals are controlled according to the prescribed system relationships.

co-processor a microprocessor that works with the central processing unit of a computer. Often the coprocessor is responsible for enacting tasks that unburden the central processing unit.

corona radiata fibers that fan out from the internal capsule to the cerebral cortex.

corpus striatum consists of the caudate nucleus and the lentiform nucleus. The lentiform nucleus consists of the putamen and the globus pallidus.

data bus often refers to the cable connection comprised of wires that transmits data information from one device to another.

decision tree logical steps that a computer or an animal uses to make judgments about interactions with the surrounding environment.

dendrite a branching neuronal protoplasmic process carrying impulses to the nerve cell body.

dendritic spine a cytoplasmic bud on the surface of a dendrite used for synaptic contact.

diencephalon consists of epithalamus, thalamus, hypothalamus and the subthalamus.

dominant hemisphere the cerebral hemisphere that is responsible for speech, usually the left hemisphere.

dopamine a neurotransmitter. Deficiency of this neurotransmitter is associated with the tremor and rigidity of Parkinsonism.

dura mater hard thick outer covering of the brain.

electrolyte a substance which when dissolved in a specified solvent (usually water) produces a conducting medium.

electromagnetic energy is energy in the form of an electromagnetic field (radio waves, light waves, x-rays, gamma rays, etc.).

electromagnetic spectrum the total range of wavelengths or frequencies in electromagnetic waves.

electron one of the smallest known particles having electric charge. Usually exits as the part of the atom outside the nucleus. The number of electrons in an atom is equal to the number of protons in the nucleus of the atom. The rest mass of an electron is equal to 9.107×10^{-28} grams.

electron charge is the fundamental unit of electrical charge; it is the charge of one electron; it is equal to 1.60203×10^{-19} coulomb.

electron current in a wire consists of the motion of electrons through the material of the wire. A current of 1 amp corresponds to the passage of 6.24 billion billion electrons in one second.

electronic 1. of or pertaining to devices, circuits, or systems utilizing electron devices. 2. also applies to the action of electrons and holes in semiconducting devices, such as transistors and diodes.

electron tube is a device in which the passage of electrons through a vacuum or a gas between two electrodes tales place within a sealed envelope, usually a glass tube. Electron tubes use a cathode which acts as a source of electrons and an anode, a more positive electrode that attracts electrons, the output of which establishes a current flow. A diode contains only a cathode and an anode. Other vacuum tubes may contain a grid electrode that is capable of producing a potential that can affect the flow of electrons streaming from the cathode to the anode. The grid electrode that is usually positioned between the cathode and anode and the magnitude of the current flow is very sensitive to the potentials imposed on the grid electrode.

ENIAC (Electronic Numerical Integrator Analyzer and Computer) was the first all-electronic computer, built between 1943 and 1945 to produce missile trajectory tables.

fasciculus a bundle of nerve fibers with in the CNS.

GABA an inhibitory neurotransmitter.

GIGO garbage in garbage out. A phrase used by computer programmers that identifies that a software program is only as good as the time and energy spent in writing and debugging the program.

glia abbreviated form of neuroglia.

glutamate an excitatory neurotransmitter.

hindbrain comprised of the pons, cerebellum, and medulla.

hippocampal formation curved band of archipallium located in the temporal lobe between the choroidal fissure and the parahippocampal gyrus responsible for memory and processing of new information.

homonymous same visual field of both eyes.

HTML (hypertext markup language) the coding language used to create the look of documents posted on the world wide web.

http part of a URL (universal resource locator) that identifies locations on the Internet that use HTML to construct web pages.

hub a device through which several computers connect on a network.

Immunology the study of the human body's immune or defense system.

internal capsule white matter between the caudate nucleus and diencephalon medially and the lentiform nucleus laterally; continuous rostrally with corona radiata and caudally with cerebral crus.

Internet a worldwide network of computers with millions of users that are linked to facilitate the exchange of information, data, conversation.

LAN (local area network) a self-contained network that may connect to the **Internet** that usually resides in a single office or building.

lemniscus a secondary sensory tract ascending through the brainstem to the thalamus.

lentiform nucleus consists of the putamen and the globus pallidus.

limbic system cortical and subcortical structures that influence behavior and autonomic responses chiefly through the hypothalamus, includes the limbic lobe, amygdaloid nucleus, hippocampal formation, septal region, and hypothalamus. Some include the anterior thalamus.

link text or graphics on a Web page that when clicked or activated, refers or directs the user to other Web pages.

local area network (LAN) a self-contained network usually in an office or home that may or may not be connected to the Internet.

mechanoreceptor in the human nervous system a receptor that is excited by its distortion due to touch, pressure, muscle or tendon stretch, etc.

medial lemniscus tract located medially in the medulla, ventrally in the pontine tegmentum and dorsolaterally in the midbrain tegmentum, carries touch, pressure, and position sense impulses from the contralateral gracile and cuneate nuclei to the ventral posteriorlateral nucleus of the thalamus.

Meissner corpuscle encapsulated tactile receptor in dermal papilla.

melanin dark brown or black pigment found in the cytoplasm of neurons in some nuclei (such as the substantia nigra, locus ceruleus).

microchip a sheet of silicon dioxide on which microscopic electric circuits have been etched.

microprocessor is silicon wafer device that is the brains of a computer that contains a collection of transistors that manipulate data in the form of binary bits.

middle cerebellar peduncle fiber bundle connecting the cerebellum and the pons.

neospinothalamic system the newer spinothalamic system which carries fast pain to the ventral posterolateral thalamic nucleus; the peripheral fibers arise from the marginal neurons in the dorsal horn of the spinal gray mater.

neuroepithelium epithelial cells (skin cells) that serve as the special receptors in the auditory (sound or hearing), vestibular (balance), olfactory,(smell) and gustatory (taste) systems.

neuroglia (also glia) non-neuronal support cells of the CNS; ten times more numerous than neurons; four types: astrocytes, oligodendrocytes, microglia, and ependymal cells.

neurology the study of the brain and the nervous system and the disease states that affect it.

neurocomputerology or neurocomputer science the study of the brain and the nervous system anatomy and function, and how the neuronetwork functions like a computer.

neurocomputerologist one who studies the brain and its anatomy in terms of how a computer functions.

neurotransmitter any specific chemical agent released by a presynaptic cell on excitation, which crosses the synaptic cleft (space) to stimulate or inhibit the postsynaptic cell.

nociceptor a pain receptor stimulated by actual tissue injury or anticipated injury.

oligodendrocytes neuroglial cells with small electron-dense oval nuclei and scanty cytoplasm that form the myelin sheath of the CNS.

operating system the software that serves as a bridge between the computer hardware and the application software in a computer.

oxytocin a hormone secreted by the hypothalamus that stimulates contraction of smooth muscle in the pregnant uterus and ducts of the mammary glands (breasts). May act as an important neurotransmitter to stimulate a woman's sexual interest in a man.

paleospinothalamic system older spinothalamic system which carries slow pain to a broader area including the reticular formation and the intralaminar thalamic nucleus.

Papez circuit neural circuit concerned with short-term memory and learning and thought to be reverberating; it includes the hippocampus, fornix, mamillary bodies, mamillothalamic tract, anterior thalamic nucleus, cingulate gyrus, cingulum, and parahippocampal gyrus.

Parkinson disease neurologic syndrome characterized by tremors at rest and rigidity ascribed to lesions of the substantia nigra.

projections fibers axons that connect the cerebral cortex with subcortical neurons.

propriospinal neurons spinal cord cells whose axons make up the fasciculi proprii adjacent to the gray matter of the brain.

reverberating is to repeat, to resound, to reecho, as in a succession of echos.

scotopic vision when the eye is dark adapted.

search engine a software program that searches documents located on the Internet for keywords or phrases entered by a person browsing the Internet. A search engine provides a list of sites related to the topic the user is searching for.

semiconducting material a material that acts like a semiconductor. The most commonly used semiconducting materials are silicon and germanium. Other semiconducting materials include selenium, cuprous oxide, lead telluride, lead sulfide, cadmium sulfide, and silicon carbide.

semiconductor a solid or liquid electronic conductor, with resistivity intermediate between that of metals and that of insulators, in which the electrical charge carrier concentration increases with increasing temperature over a range of temperature, over the most practical temperature range, the resistance has a negative temperature coefficient.

Certain semiconductors possess two types of carriers, namely, negative electrons and positive holes.

semiconductor device an electron device in which the characteristic distinguishing electronic conduction takes place within the semiconductor. This term usually applies silicon devices such as transistors, microprocessors, memory chips, and central processing units.

septal region that portion of the human brain that is associated with reward or pleasurable feelings, it is located in the limbic system area anterior and lateral to the lamina terminalis and includes the subcallosal area and the septal nuclei.

server a computer in a network that provides files and services for other computers termed 'clients'. A file server provides file storage that is accessible to the client computers. A print server provides printing capacity to the clients on the network. A mail server provides mail services between the clients on a network and the Internet.

signal a visual, audible or other indication used to convey information. The intelligence or message to be conveyed over a communication system.

slow pain dull, burning pain that is diffuse rather than localized, resulting from tissue injury.

somatosensory system pertains to general somatic senses: somatic pain and temperature, touch, vibration, and limb position and motion sensibility.

spasticity condition of increased muscle tone and exaggerated tendon reflexes.

spiral organ sensory end organ for hearing found in the cochlear duct of internal ear.

split brain a brain in which the corpus callosum and anterior and hippocampal commissures have been severed down the medial plane.

stereognosis ability to recognize an object by touch.

substantia nigra is one of the basal ganglia and is a pigmented nuclear mass located in the midbrain which when malfunctions is associated with tremors and rigidity of Parkinson's disease.

subthalamic nucleus is one of the basal ganglia, and is a nuclear mass located in the subthalamus which when malfunctions is associated with ballismus.

subthalamus also called the ventral thalamus, is comprised of the sub-thalamic nucleus, zona incerta and prerubral field, it is part of the diencephalon found between the thalamus dorsally, the cerebral peduncle ventrally, and the hypothalamus medially.

sulcus refers to a groove, furrow, depression or fissure.

superior cerebellar peduncle fiber bundle connecting the cerebellum and the midbrain.

suprachiasmatic nucleus (SCN) located in the anterior hypothalamus, generates the internal circadian rhythm.

sympathetic that division of the autonomic nervous system that plays a role in the preparation for emergency situations.

synapse site of functional contact between neurons where impulses pass from one neuron to another.

synaptic cleft the space between where two neurons come together or connect.

TCP/IP (transmission control protocol/Internet protocol) a universal standard collection of protocols use to connect servers on the Internet to facilitate the exchange of data.

transformer an electric device, without moving parts, which by electromagnetic induction transforms electric energy from one circuit to another circuit at the same frequency, usually with changed values of voltage and current (amperes).

transistor an active semiconductor device with three or more electrodes, usually a base, emitter, and a collector.

uncinate fasciculus association nerve bundle connecting the frontal and temporal lobes of the brain.

universal resource locator (URL) an address of documents and other resources on the world wide web.

vacuum tube a device in which the passage of electrons through a vacuum between two electrodes takes place within a sealed envelope, usually a glass tube. Electron tubes use a cathode, which acts as a source of electrons and an anode, a more positive electrode that attracts electrons, the output of which establishes a current flow. A diode contains only a cathode and an anode. Other vacuum tubes may contain a grid electrode that is capable of producing a potential that can affect the flow of electrons streaming from the cathode to the anode. The grid electrode that is usually positioned between the cathode and anode and the magnitude of the current flow is very sensitive to the potentials imposed on the grid electrode.

vasopressin (antidiuretic hormone (ADH)) produced by the pituitary gland, has an antidiuretic effect and a pressor effect that elevates blood pressure. In men may act as an important neurotransmitter to stimulate a man's sexual interest in a woman.

vermis the midline portion of the cerebellum; it connects to the fastigial nucleus and affects the vestibular nuclei for equilibrium and eye movements.

Wernicke's area posterior part of the superior temporal gyrus of the dominant hemisphere, that functions as a receptive speech center in the human brain.

wide area network (WAN) a single network that extends beyond the boundaries of an office, building or home.

virus 1. computer code that has the ability to replicate, hide, watch for a triggering event, and deliver a destructive or prankish action on a computer or computer network. 2. An organism comprised of a protein outer shell surrounding genetic material in its inner core, that invades a larger organism, inserts itself into the genetic material of the organism and in using the organism's resources replicates itself, then escapes the organism to infect others.

World Wide Web (WWW) a loose confederation of Internet servers that utilize HTML formatting to facilitate communication with each other connected on the net.

LIST OF COMMON ABBREVIATIONS

ADC analog-to-digital converter

AGP accelerated graphics port

ALU arthritic logic unit

ASCII American Standard Code for Information Interchange

BIOS basic input/output system

BOSS brain operating system software

CD-R compact disc recordable

CD-ROM compact disc read only memory

CD-RW compact disc rewritable

CMOS complementary metal-oxide semiconductor

CNS central nervous system

CPU central processing unit

CRT cathode ray tube

DAC digital-to-analog converter

DAT digital audio tape

DIMM dual in-line memory modules

DLL dynamic linked library

DNS domain name system

DOS disk operating system

DRAM dynamic random access memory

DSP digital signal processor

DVD digital versatile disk

EPROM erasable, programmable, read-only memory

FAT file allocation table

GHz gigahertz measurement in billions, the number of times something oscillates or vibrates.

HTML hypertext markup language

IPS identity police service

IDE integrated drive electronics

LAN local area network

LCD liquid crystal display

MHZ megahertz measurement in millions, the number of times something oscillates or vibrates.

MS-DOS Microsoft disc operating system

NIC network interface card

PC personal computer

PCI peripheral component interconnect

POP post office protocol

POST power-on self-test

RAM random access memory

ROM read only memory

SCN suprachiasmatic nucleus

SRAM static random access memory

SDRAM synchronous dynamic random access memory

SMTP simple mail transfer protocol

TCP/IP transmission control protocol/Internet protocol

UPS uninterruptible power supply

URL universal resource locator

USB universal serial bus

VDSL very high-speed digital subscriber line

VFAT virtual file allocation table

VRAM video random access memory

WWW world wide web

XML extensible markup language

FURTHER READING

ATLAS OF NEUROANATOMY, Joseph J. Warner, Butterworth Heinemann, 2001.

ATLAS OF THE UNIVERSE, Patrick Moore, Cambridge University Press, 1998.

BASIC CLINICAL NEUROANATOMY, Paul A. Young & Paul H. Young, Lippincott Williams & Wilkins, 1997.

BASIC IMMUNOLOGY, Jacqueline Sharon, Williams & Wilkins, 1998.

BIOLOGY, Karen Arms & Pamela S. Camp, W.B. Saunders Company, 1979.

BIOMEDICAL INSTRUMENTATION AND MEASUREMENTS, 2nd ed, Leslie Cromwell & Fred J. Weibell & Erich A. Pfeiffer, Prentice-Hall Inc, 1980.

THE CIBA COLLECTION OF MEDICAL ILLUSTRATIONS, Vol 1, NERVOUS SYSTEM, Frank H. Netter, CIBA, 1986.

EARTH PRO, Anthony Scheiber, iUniverse, 2000.

ENCYCLOPEDIC DICTIONARY OF ELECTRONICS AND NUCLEAR ENGINEERING, Robert I. Sarbacher, Prentice-Hall Inc., 1959.

EVOLUTION, Carl Zimmer, Harper Collins Publishers, 2001.

HARRISON'S PRINCIPLES OF INTERNAL MEDICINE, 12th ed, McGraw-Hill Inc., 1991.

HOW COMPUTERS WORK, 6th ed, Ron White, Que Corp, 2002.

IMMUNOLOGY, 5th ed, Ivan Roitt & Jonathan Brostoff & David Male, Mosby, 1998.

LEADERSHIP SECRETS OF ATTILA THE HUN, Wess Roberts, Warner Books, 1987.

MANTER AND GANTZ'S ESSENTIALS OF CLINICAL NEUROANATOMY AND NEUROPHYSIOLOGY, edition 9, Sid Gilman, Sarah Winans Newman, F.A. Davis Company, 1996.

THE PLANETS, David McNab & James Younger, Yale University Press, 1999.

PRINCIPLES OF NEUROLOGY, Maurice Victor & Allan H. Ropper, McGraw Hill, 2001.

TEXTBOOK OF CLINICAL NEUROLOGY, Christopher G. Goez & Eric J. Pappert, W.B. Saunders Company, 1999.

UPGRADING AND REPAIRING PCS, 13th ed, Scott Mueller, Que Corp, 2002.

FINAL NOTE

A respected professor of mine once remarked, "The most interesting medical research occurs after nine o'clock at night."

At the time, being much younger, I thought the guy was quite crazy. Well, time has a way of smoothing away rough surfaces, opening your eyes, and changing perspectives. I apologize for the mistakes, possible omissions, outlandish assumptions that lurk in the pages of this book.

After all the day's patients had been seen in the clinic, and all the notes written, and all the bills paid, and the boys' homework done and tucking them in for the night, and after the house was picked up and the dog was walked and the phase of the moon observed, finally came time to write the text of this book. My appreciation for the wisdom of my professor, glows inside me every night...and I hope it will glow inside of you and inspire you to do wondrously great things.

Good night.

0-595-25293-1

www.ingramcontent.com/pod-product-compliance
Lightning Source LLC
Chambersburg PA
CBHW020721180526
45163CB00001B/54